A specification manual for...
Johns-Manville
BUILT-UP ROOFS

Johns-Manville recognizes the need for Built-Up Roofing Specifications to be adjusted for various climatic conditions.

Therefore we have divided the United States into Three Geographic Regions shown below.

If there are limitations for use of any Roofing Specification in this manual such regional limitation is indicated in the Roof Finder Index (page 34) and on the individual specification.

For _____

Presented by _____

Issued Jan. 1, 1976

Table of Contents

General	1
Roof Guarantee	2
A Statement of Policy	4
Test Cuts	6
Approved Roofing Contractors	6
Roofing Materials	7
Roof Decks	17
Roofing Specifications	31
Roof Insulation Specifications	203

General

This manual has been prepared as a reference for Architects, Engineers, and Roofing Contractors. It presents a range of roofs for different roof inclines and types of roof deck.

The user is urged to familiarize himself with all the material included before selecting a particular roof specification.

Johns-Manville has been manufacturing built-up roofing materials for more than 100 years. The information and specifications contained herein are based upon that manufacturing background and extensive field experience and is supplemented by thorough and continuing research. Necessarily the specifications are designed for normal installation conditions. For unusual conditions, a Johns-Manville roofing specialist should be contacted.

If non-standard sub-strates or unusual shapes such as domes, etc., are to be roofed, prior approval must be obtained from Johns-Manville as to the roofing specification and method of application to be used.

A Johns-Manville Sales Representative, thoroughly oriented and trained in built-up roofing, is available to you. They have technical support from District Engineers who in turn have support from highly trained Research and Engineering personnel for consultation and technical assistance.

All information and specifications for Built-Up Roofing contained in this manual supersede any prior published data by J-M on this subject.

The physical properties of Johns-Manville Built-Up Roofing and Roof Insulation represent typical, average values obtained in accordance with accepted test methods and are subject to normal manufacturing variations. They are supplied as a technical service and are subject to change without notice. Check the J-M district office to assure current information.

JM Johns-Manville Roof Guarantee

Guarantee No:
Date:

Applied by:
Location of Building:
Name of Owner:
Owner's Address:
Name of Building:
Date of completion of roofing:
Approximate area guaranteed:
Amount of Guarantee:

Roof Specification No.

Johns-Manville Sales Corporation, a corporation of the State of Delaware, with an office at Greenwood Plaza, Denver, Colorado, 80217 (sometimes hereinafter called "J-M"), guarantees to the original owner named above that under and subject to all the conditions herein set forth, for a period of _____ years from the date of completion of the roof, J-M will at its own cost and expense, to the extent of an aggregate cumulative expenditure not exceeding the amount of the Guarantee shown above, make or cause to be made any repairs that may become necessary to maintain the "J-M" roof, exclusive of flashings, metal work, a vapor retarder and insulation in a watertight condition.

This Guarantee is made under and subject to the following conditions:

1. **This Guarantee is given and accepted in lieu of all other liabilities or warranties on the part of any and all Johns-Manville Corporations, express or implied, in fact or in law, including without limitation the warranty of merchantability and the warranty of fitness for a particular purpose.**

2. **The presence of standing water on the roof surface will invalidate this guarantee.**

3. J-M's sole responsibility and liability under this Guarantee and owner's exclusive remedy hereunder is specifically limited to the repairs that may become necessary to maintain the J-M roof in a watertight condition subject to all of the conditions herein set forth. No Johns-Manville Corporation shall have any other liability or responsibility with respect to this Guarantee, and any liability for incidental or consequential damages of any nature whatsoever, including without limitation loss of profits, is hereby excluded.

4. The owner of the roof will notify J-M if repairs covered hereunder are required. The notice shall be given to J-M in writing and shall be delivered at the office of J-M by certified or registered mail within sixty (60) days after the necessity of such repairs is or should have been discovered. Failure to give such notice as aforesaid shall negate and void all responsibility of J-M under this Guarantee.

5. All responsibility and obligation hereunder shall terminate and cease on any of the following events:

   A. If a sprinkler system, water or air cooling equipment, radio or television aerial or aerials, chimneys for steam, water tower or other addition shall be installed on the roof after the completion of the roof, unless J-M shall first be notified of the making of the necessary roofing application with respect thereto and the materials to be used to join such proposed installation to the roof and the method of application of such materials and structures.

   B. If any repair work shall be done other than under the supervision of or subject to the inspection and approval of J-M;

   C. If the roof is damaged by natural disasters, i.e. windstorm, hail, flood, hurricane, lightning, tornado, earthquake or other phenomena of the elements;

   D. If the roof has been subject to misuse, negligence or accident in any way, or

   E. The owner fails to give notice or to comply strictly with each and every term or condition specified herein to be performed by him.

F. If the roof has been subject to test cuts that have not been approved by J-M.

G. If the owner fails to make repairs for which he is responsible within a 90-day period after the discovery of need for repair is made or should be made in the exercise of reasonable care.

6. No suit shall be brought under this Guarantee later than one year after the necessity for the repairs is or should have been discovered.

7. This Guarantee shall accrue only to the original owner named above. It shall not accrue to the benefit of any tenant, purchaser, successor or assign of the original named owner.

Leaks from the following causes, except when caused by those exclusions set forth below, are covered by this Guarantee:

1. Natural deterioration of membrane.
2. Bare spots.
3. Slippage.
4. Fish mouths.
5. Ridges.
6. Splits.
7. Buckles and wrinkles.
8. Thermal shock.
9. Workmanship in application of roofing membrane.
10. Workmanship in application of base flashing (only if endorsed).
11. Natural deterioration of base flashing (only if endorsed).
12. Slippage of roofing membrane.
13. Slippage of base flashing (only if endorsed).

EXCLUSIONS —

By way of illustration and example and not limitation, the following types of damage, change and repairs are excluded from this Guarantee:

1. Roof maintenance for corrections of conditions other than leaks.
2. Natural disasters, i.e., windstorm, hail, flood, hurricane, lightning, tornado, earthquake or other phenomena of the elements.
3. Structural defects or failures.
4. Damage to building or its contents.
5. Changes in building usage unless approved in writing in advance.
6. Damage resulting from any new installation on, through, or adjacent to the roofing membrane.
7. Repairs or other applications to the membrane or base flashing after date of completion unless performed in a manner acceptable to J-M.
8. Any material used as a roof base or insulation over which a J-M roof membrane is applied.

In Witness Whereof: Johns-Manville has caused this Guarantee to be duly executed the date set forth above.

ATTEST:

JOHNS-MANVILLE SALES CORPORATION

_____
Assistant Secretary

By: _____
Attorney-in-Fact

JOHNS-MANVILLE SALES CORPORATION ("J-M") FLASHING ENDORSEMENT

THIS ENDORSEMENT is attached to and made part of J-M Roof Guarantee Certificate of Coverage No. _____ years from the date J-M guarantees under the conditions set forth in the above-mentioned certificate that for a period of _____ years from the date of the completion of the roof described above it will at its own expense within and as part of the total aggregate cumulative amount of Guarantee above set forth make any repairs that may become necessary to maintain in a watertight condition _____ lineal feet of Johns-Manville Flashing Material & _____ lineal feet of Expand-O-Flash Expansion Joint Covers used in conjunction with this J-M Roof. The Guarantee does not cover metal work of any kind used in conjunction with flashing nor any damage occasioned by defects in, or infiltration of moisture through, the walls, coping or building structure.

ATTEST:

JOHNS-MANVILLE SALES CORPORATION

_____
Assistant Secretary

By: _____
Attorney-in-Fact

Johns-Manville Roofing Guarantee

Roofing guarantees are available from Johns-Manville for most built-up roofing systems published in this Manual. See Roof Finder Index for guarantee periods available.

Flashing endorsement periods are noted in individual specifications. Guarantees are the owners' assurance that the installed roofing systems, under normal wear and tear, will be watertight for the total specified number of years with a varying guarantee up to $20 per square. Backed by well over a century of experience in the manufacture and supervision of installations of built-up roofing, J-M is aware that providing excellent field service to back up its well-engineered and proven product is vital to insure the best possible roof performance. There are many Johns-Manville engineers and roofing specialists throughout the nation who are available to assist in the installation and servicing of all Johns-Manville guaranteed roofing systems.

Johns-Manville manufactures roofing materials. It does not practice architecture or engineering. The roofing systems included in this manual result in satisfactory installations when properly applied. J-M makes no implied warranties whatever, as to materials of our manufacture or performance of the systems described herein. When J-M Roof Guarantee is given, the express warranties are only those specifically enumerated therein. No agent, salesman, representative nor employee is empowered to change, alter, or amend this provision unless it is done in writing by this headquarters office in Denver, Colorado.

A statement of policy

The information and specifications contained herein are intended to assist in selecting appropriate roofing specifications. Basic design requirements in structures are the responsibility of the building designer, therefore, the roofing systems manufacturer's recommendations and requirements should focus on the roofing system (vapor retarder, insulation, membrane) and details of construction which directly affect its application and performance. Johns-Manville does not assume responsibility for decisions as to when and where vapor retarder systems or special attachment procedures for abnormal wind factors are advisable. When these decisions are factors, the recommendations and procedures outlined herein should be considered. The basic recommendations and suggestions are provided only to assist the designer; Johns-Manville will not assume any responsibility for building structural design adequacy or performance of deck or other elements of the building system not included in the roofing system. Johns-Manville will not be responsible for any guarantee for failure of the roofing system due to structural defects, damage of other building trades, or for failures due to errors in design of any building element.

Because all of the factors creating abnormal wind conditions on a roof that cannot be anticipated by a roofing manufacturer, J-M cannot accept wind damage liability.

When a roofing system has been applied by a Johns-Manville Approved Roofing Contractor using Johns-Manville materials *to the satisfaction of Johns-Manville*, a Roofing guarantee may be issued to the Roofing Contractor for a charge based upon the fee scheduled in effect at time of completion of the roof. Johns-Manville will not accept Notices of Award of Contract which indicate that the Owner or Architect has the option to accept or reject the guarantee upon completion of the roof.

Notice of award must be received and accepted, by the J-M District Engineer before roof application commences.
The guarantee cannot be cancelled by the owner once it has been issued, except with the agreement of Johns-Manville.

Johns-Manville reserves the right to change or modify at its discretion and without prior notice any of the information, recommendations, specifications, or the terms of the guarantee.

If conditions are observed on the job site that do not conform to the requirements for the application of a J-M guaranteed roof, our representative will advise the Roofing Contractor or the Architect's representative of such conditions.

No guarantee or product warranty will be issued on any built-up roof assembly over the following:

Roof Areas less than 50 squares.
Cold Storage Buildings.
Private Residences.
Storage Silos and Heated Tanks.
Structures outside the contiguous states of the United States.
Conduit or Piping over the Roof Deck under Built-Up Roofing, except where conduit or piping is installed in channels of deck permitting full and uniform thickness of insulation over entire deck surface.
Plastic Foam insulations except those approved by Special Bulletin.
Lightweight insulating poured fills unless Ventsulation Felt is used in accordance with Johns-Manville specifications.
Re-Roofing over old insulated roofs (with or without old roofing removed) unless Ventsulation Felt is used under new assembly.
A roof or any area of a roof that is not readily accessible for inspection.

When no guarantee is purchased, Johns-Manville will not write or sign letters; (a) stating that our representatives have examined plans, details or specifications which are acceptable for the receipt of Johns-Manville roofing and flashing systems, (b) stating that a roof has been applied according to Johns-Manville specifications or recommendations for either guaranteed or non-guaranteed jobs, (c) or verbally issue any guarantee or product warranty other than the published product warranty.

Only roof installations that are to be guaranteed will be inspected by the J-M Sales Representative, provided one half the guarantee charge has been paid. Johns-Manville reserves the right, however, to waive inspections on guaranteed roofs when, in its opinion, inspection is not necessary, or location is remote from a company location. In such cases, the owner or designer may request inspection for which an additional fee will be charged.

Test cuts

Johns-Manville opposes taking test cuts from new roofs to judge roof compliance with specifications. Failure of small samples to exactly meet total specified weight is often the basis for entire roof rejection by the owner or architect, regardless of extenuating circumstances. No recognition is given to existent variables beyond control of the roofing mechanic's using generally accepted techniques and equipment. Good distribution of bitumen is far more important to serviceability of roofing membranes than total weight of bitumen. Roofers may be forced to use excessive bitumen which can result in sliding complaints even on very low slopes.

When test cuts are made in either old or new roofs, they should be 4" x 42" cuts at right angles to the length of felts. Such samples are more meaningful than 12" x 12" samples, since they reveal whether the base felt has been properly lapped and ply felts laid with proper exposure. Allowable weight tolerance of test cut components is 15% plus or minus.

As test cut areas are of considerable size, care should be taken that when being repaired the same materials and techniques are employed as used in constructing the original roofing membrane.

The patched area will be weaker than the original roof, since it is impossible to restore completely the integrity of the original membrane.

Johns-Manville reserves the right to cancel a Roof Guarantee if the roof has been subject to test cuts that have not been approved by Johns-Manville.

Approved roofing contractors

The designation Johns-Manville Approved Roofer solely identifies a contractor as being eligible to apply for a guarantee and in no way designates a contractor as being an agent for Johns-Manville.

Roofing Materials

Asbestos Felts 9

Organic Felts 10

Roofing Bitumens 10

Gravel or Slag Surfacing 12

Cold Cements and Coatings and Primer 12

Do's and Don'ts
for Storage and Handling of Materials 14

Johns-Manville
Roofing materials

NOTE: If any or all components of a built-up roof are required to comply with a Federal or ASTM specification, it must be so noted on any order for such materials placed with J-M. Such Federal and ASTM specifications usually require certification that materials supplied meet the applicable specifications and such certification also requires tests that can only be conducted at the manufacturers plant.

Asbestos Felts

Asbestos Felts used in built-up roofing are of several different types. All have as the principal component a felt composed primarily of asbestos fiber, a non-rotting, non-wicking, inorganic mineral fiber. Depending on the end use of the product, it may be a saturated felt or it may be a saturated felt with a coating asphalt added to both sides for base felts and other uses, or coating may be added and on one side granules embedded to provide a decorative effect and weather protection.

In all asbestos felt products the inorganic character of the felt provides longer life, greater resistance to weathering, and lower maintenance costs. Johns-Manville Asbestos Felt Roofing specifications are available with smooth, gravel, or mineral surfacing.

Following is a compilation of the Asbestos Felt Products referred to in this manual.

Product	Sq. Per Roll	Approx. Weight Per Roll	Applicable Specifications Federal	ASTM	Navy
Centurian Base Felt	Two	50 lbs.	HH-R-590 Type II, Class A (Unperforated)	D-250	TS-07510
Coated Asbestos Base Felt	One And Two Region 3 Only	43 lbs. 86 lbs.	—	D-3378	—
Asbestos Finishing Felt	Four Three Region 3 Only	60 lbs. 45 lbs.	HH-R-590, Type I, Class A (Perforated) Type II, Class A (Unperforated)	D-250	TS-07510
Reinforced Base Flashing	12"-⅓ 18"-½ 36"-1	22 lbs. 33 lbs. 65 lbs.	—	—	—
GlasKap Mineral Surface Cap Sheet	One	72 lbs.	SS-R-630 d Class 3	—	—
Ventsulation Felt	One	70 lbs.	—	—	—
Cold Application Felt	Two	70 lbs.	—	—	—
Special Coated Asbestos Base Felt Region 3 Only	One	55 lbs.	HH-F-182 Class B	D-655 Type 501b	TS-07510
Blue Chip Asbestos Finishing Felt	Three	60 lbs.	HH-R-590 Type I, Class A Perforated	D-250	TS-07510
Asbestogard Felt	Five	45 lbs.	—	—	—
Flexstone Mineral Surface Cap Sheet	One Region 3 Only	77 lbs.	SS-R-630 d Class 2	—	—

Organic Felts

Organic felts are made of fibrous organic material. Since they are subject to deterioration by oxidation and to wicking, they must be saturated with bituminous saturant.

In a built-up roof organic felts must be flooded with a heavy coat of asphalt and gravel embedded therein to prevent rapid weathering and rotting. An organic felt roof must be fairly heavy (usually about 500 to 600 lbs. per square). Small breaks and leaks are difficult to find and repair because of the presence of the overlying gravel layer.

Following is a compilation of the organic felt products referred to in this manual:

Product	Sq. Per Roll	Approx. Weight Per Roll	Applicable Specification			
			Federal	ASTM	U.S. Navy	Others
#15 Asphalt Saturated	Four Three (Region 3 Only)	60 lbs. 45 lbs.	HH-R-595-B Type 15A Style B	D-226 UPON REQUEST	TS-07510	A.R.E.A. (Unperforated) Comm. 29, sec. 210.3
Planet Base	1½ Two - (Region 3 Only)	64 lbs. 86 lbs.	SS-R-501D Type 1	D-2626 Type 1	TS-07510	
ACCESSORY ITEMS							
Sheathing Paper	Five	20 lbs.	—	—		

Roofing Bitumens

Roofing asphalts come in many grades. In general they are grade-specified by softening point. The slope of the roof governs the grade to be used. It is very important that the proper grade as shown in the specification be used.

While asphalts are not as susceptible to damage from overheating as Coal Tar derivative material, it is most important that precautions be taken against kettle overheating. Overheating even for short periods can cause "cracking" or molecular change in the asphalt (a drop in softening point and slight oiliness is a symptom) or if the overheating is more gradual the asphalt may be "aged" to the extent that premature failure will result on the roof. An indication of the latter is increase in softening point and possibly a hard, brittle surface with early "alligatoring".

Since asphalts are thermoplastic, their viscosity varies with temperature. Application temperatures are specified in the temperature range which will permit an adequate application of material by mop or machine.

Cutting back, adulterating or mixing of asphalts with any other material is not to be allowed.

Application and kettle temperatures for various asphalts, shown in the table below, cover the range for most asphalts presently used. However, as conditions can vary, heating temperatures may have to be adjusted to obtain the recommended mopping quantity of 23 lbs./square Kettle temperatures of 500°F or more for Type III & IV Asphalt should not be maintained for more than four hours, without danger of seriously damaging the asphalt. However, some asphalts may have a flash point that will not tolerate this temperature. Do not exceed this Flash Point or the recommendation of the vendor.

Use of insulated buckets and hoses is recommended for cold weather application.

Johns-Manville Roofing Asphalts specified for use in constructing built-up roof coverings are tested in accordance with ASTM-D-312. Shown below are the values as given in ASTM-D-312 and for comparison those of Johns-Manville specifications, which in every instance meet or are more restrictive than ASTM.

PHYSICAL PROPERTIES

Asphalt Type	Softening Point — °F Min.	Softening Point — °F Max.	Flash Point C.O.C. °F	Penetration 32° — 60 Sec. 200 g. Min.	Penetration 32° — 60 Sec. 200 g. Max.	Penetration 77° — 5 Sec. 100 g. Min.	Penetration 77° — 5 Sec. 100 g. Max.	Penetration 115° — 5 Sec. 50 g. Min.	Penetration 115° — 5 Sec. 50 g. Max.	Ductility .77°	Slide — Inches 6 Hrs. @ 130°F Min.	Slide — Inches 6 Hrs. @ 130°F Max.	Recommended Heating and Application Temperatures Heating	Recommended Heating and Application Temperatures Appl.
ASTM-D-312 Type I	135	150	437	3	—	18	60	90	180	10	—	—	425°F	275°F to 350°F
J-M Aquadam	140	150	500	3	—	18	35	90	150	12	2.2	4.2		
ASTM-D-312 Type III	180	200	437	6	—	15	35	—	90	3	—	—	450°F	350°F to 425°F
J-M 190 Grade	185	195	500	10	—	20	35	—	60	3	—	1.4		
ASTM-D-312 Type IV	205	225	437	6	—	12	25	—	75	1.5	—	—	475°F	375°F to 450°F
J-M 220 Grade	215	225	500	10	—	12	25	—	75	2	—	.5		

Gravel or slag surfacing

Any gravel or slag that is damp must be dried before using to prevent foaming of bitumen as gravel or slag is applied. In cold weather, if any difficulty is experienced in obtaining proper embedment in the bitumen, the gravel or slag shall be heated immediately before application.

Johns-Manville will approve the use of clean slag or gravel meeting ASTM Specification D1863-64.

ASTM Specification D1863-64 covers aggregates specified both for use in road construction and built-up roofing and should be commercially available throughout the country.

Other surfacing material used in place of gravel or slag should be fairly cubical in shape, non-water absorbent, hard and opaque, and of such size and nature as to result in firm embedment in the asphalt.

Cold cements and coatings and primer

Several asphaltic cements and coatings are designed for specific purposes in connection with built-up roofing. These are used primarily as primer, as surfacing for smooth-surface asbestos roofs, or as flashing cement. With certain types of felts a cold cement may be used to cement the felts together or secure them to decks. Following is a brief description of some of these:

Product	Use	Application Method	Applicable Specification
Cold Application Cement	Cement Plies of Cold Application Felt	Brush	—
Asbestile	Flashing Cement	Trowel	ASTM-D-2822
Aquapatch	Patching on Wet or Dry Surfaces	Trowel	—
Topgard Type B – Fibrated	Emulsion Roof Coating Roof Slopes ½" & Above	Brush or Spray	ASTM-D 1227 Type I Mil-R-3472 Clay Type
Topgard Type C	Cut Back Roof Coating — Unfibrated	Brush or Spray	—
Topgard Type F — Fibrated	Cut Back Roof Coating	Brush or Spray	SS-A-694-D
Fibrated Aluminum Roof Coating	Cut Back Roof Coating — Reflective	Brush or Spray	—
Concrete Primer	Priming Non-Wood Surfaces	Brush or Spray	F.S.-SS-A-701 ASTM-D-41
Industrial Roof Cement	All-Purpose Roofing Cement	Trowel	F.S.-SS-C-153 (Type I) ASTM-D-2822
Asbestogard Adhesive	Class I F-M Construction	Roller	—
CeramaKap Adhesive	Cold Application Roof Systems	Brush or Spray	—

For Coverage See Application Instructions

The bituminous surfacing on a new built-up roof will experience considerable movement. Because of this, colored coatings such as Fibrated Aluminum Roof Coating should not be applied until the roof surface has weathered at least one summer. Aquadam never completely stabilizes as it maintains self-healing and flow properties throughout its life.

Roof surfaces (½" per foot slope & above) to receive Topgard Type-B must be clean, dry, & free from dust or dirt & primed prior to application of this coating.

If a smooth surface roof is to receive Fibrated Aluminum Roof Coating, and an additional surfacing is desired, it should be surfaced with Topgard Type-B rather than Aquadam, prior to the application of the Aluminum Roof Coating.

No colored roof coating will resist standing water. Valleys and low spots should be top poured with Aquadam and surfaced with gravel or reflective aggregate such as marble chips.

Do's and Don'ts
for Storage and Handling of Materials

Do's

- Always stand roll goods on end.
- Store roll goods in a dry place protected from sun and weather.
- Roll out and cut mineral surface and asphalt coated roofing into sheets and allow them to flatten before use, especially when application is to be made in air temperatures below 50°F.
- Always place roll goods on clean floors or on platforms in such a way as to prevent damage to ends and embedment of foreign matter in ends.
- Handle roll goods with care. Dropping rolls on edges or ends damages and deforms the felt often preventing proper application.
- Store cartons and drums of roofing asphalt on end on level standing to prevent flow from containers.
- Store paper cartons of roofing asphalt protected from sun and weather to prevent carton deterioration.
- Store all roofing asphalt so it is protected from rain and snow. Moisture absorbed into the asphalt under such conditions can cause kettle foaming and poor application.
- Store all materials containing solvents in dry, cool storage with proper fire and safety precautions.
- Store all emulsions in dry storage at temperatures over 40°F.
- Heat roofing asphalts only within temperature limits prescribed by the manufacturer's specifications. Overheating can cause changes in physical properties and results in inadequate application.
- Replace lids tightly on cans of material containing solvent or water based materials.
- Do Broom all Felts manually, immediately following either Felt-Layers or Hand Mopping.
- Do see note on page 29 concerning cold weather precautions.

Don'ts

- Don't leave roofing materials unprotected on the job or elsewhere.
- Don't use wet or damaged roofing materials.
- Don't overheat roofing asphalts.
- Don't store coated roll goods in air temperatures below 40°F.
- Don't apply coated roll goods when the temperature is below 40°F unless the mopping asphalt is kept at the proper temperature.
- Don't allow emulsions to freeze.
- Don't apply emulsions when air temperature is below 40°F or is expected to go below 40°F within 24 hours after application.
- Don't use emulsion which has been frozen. Freezing destroys an emulsion.
- Don't dilute an emulsion without specific instructions from the manufacturer.
- Don't leave lids off of emulsions or solvent-containing materials. Evaporation of the light ingredients makes material difficult to handle and may destroy it.
- Don't heat asphalt in kettles containing remnants of other types of bitumens.
- Don't apply asphalt, roll goods, or other asphalt materials during rain or snow or to wet surfaces.
- Don't store roll goods inside or outside at air temperatures or under such conditions that the temperature of the rolls exceeds 150°F. High temperatures can cause sticking in rolls and even deterioration of some materials.
- Don't store solvent-containing materials for prolonged periods in storage when temperatures will exceed 100°F or when cans are exposed to hot sun. Rapid solvent evaporation under such conditions can cause lids to blow off or cans to burst.
- Don't apply solvent-containing materials in confined spaces. Solvent fumes can be injurious.
- Don't install Roofing in phases. Complete installation of all plies of roofing on same day.

Roof Decks

Roof Drainage	19
Steel Decks	20
Concrete Decks	21
Pre-Cast Concrete Slabs	
Pre-Stressed Pre-Cast T or TT Long Span Slabs	
Lightweight insulating Concrete Decks Poured-in Place	
Wood Decks	23
Board Decks	
Plywood Decks	
Gypsum Decks	23
Poured Gypsum Decks	
Pre-Cast Gypsum Tile	
Structural Wood Fiber Deck Units	24
Above Deck Fills	25
Wood Nailers	26
Roof Deck and Fastener Data	27
Expansion Joints	28
Vapor Retarders	28
Roof Insulation	29

Roof Decks

The primary function of a roof deck is to provide support for the roofing membrane and perhaps roof insulation. Some decks are required to furnish inside appearance and even sound control. Our interest here is in the roof deck as a support for the roofing membrane. To perform this function the deck must be reasonably rigid, it must be reasonably smooth and free of large cracks, holes, or sharp changes in elevation of the surface; it must be designed and built with adequate slope to provide free drainage to the drains without ponding water between supporting members, and to maintain the slope for the life of the building; and it must be able to receive the roof system by some method which will hold the system securely, either by cementing or by mechanical fastening. Before roofing work is started the deck should be inspected carefully by the roofing contractor and deck contractor and the owner's representative to determine that it satisfies these conditions. Providing such a deck for the roof system is the responsibility of the designer of the building, for only he is in a position to integrate these requirements with other structural and use considerations.

Roof Drainage

A roof deck should not be designed to hold water, but should in fact be designed to shed water to the drains as fast as possible.

There are many hazards in low slope decks. A deck designed on the drawing board for ¼" per foot slope seldom gets built with the specified uniform slope, and usually ends in a deck with areas of standing water.

Standing water on roof areas is hazardous in two ways — possible seepage of moisture through the membrane into the roof system, and damage to the roof from freeze-thaw cycles during the winter. No multi-ply membrane can be applied over a large area by practical roofing methods with complete perfection.

Once even relatively small amounts of moisture get beneath the roof membrane serious damage can be caused to an insulation layer and to the membrane itself through repetitive cycles of evaporation, condensation, and freezing. Eventual damage has frequently required complete replacement of the entire system.

Ponds of water on roofs in areas subject to winter freeze and thaw create further hazards. Ice formations move constantly with temperature changes. This movement can "scrub" the roof surface to the extent that considerable physical damage is done to the membrane.

The effects of ponded water have been studied by members of the roofing industry for many years. The Built-Up Roofing Committee of the Asphalt Roofing Manufacturers Association unanimously recommends that roof design provide minimum ¼" per foot slope under such conditions that the roof drain freely throughout the life of the building.

This recommendation also received the approval and concurrence of not only Johns-Manville but the National Roofing Contractors Assoc., the Mid-West Roofing Contractors Assoc., & the American Institute of Architects, as well as other regional Roofing Contractor's Association.

Johns-Manville will not be responsible for roof damage or failure at areas where water stands or damage to the system due to standing water.

The existence of this condition will void the roof guarantee.

Steel decks

Steel decks lighter than 22 gauge are not considered as an acceptable base for a Johns-Manville Built-Up Roof.

Factory Mutual publishes the following criteria for gauge and span of steel roof decks based on rib openings of 1" (narrow rib) 1¾" (intermediate rib) and 2½" (wide rib).

Maximum Spans for 1½-in. Deep Deck With 6-inch Rib Spacing.

Gage	Uncoated Steel Minimum Thickness in.	Narrow Rib	Intermediate Rib	Wide Rib
22	0.0284	4 ft 6 in.	5 ft 0 in.	5 ft 6 in.
20	0.0344	5 ft 0 in.	5 ft 6 in.	6 ft 0 in.
18	0.0449	6 ft 0 in.	6 ft 6 in.	7 ft 0 in.
16	0.0568	6 ft 6 in.		8 ft 0 in.

Above spans apply where deck is continuous over three or more supports. For single-span conditions, use 85% of the above spans. Spans are based on the uncoated steel thickness.

Johns-Manville subscribes to this criteria and we strongly recommend that architects and designers not exceed these limits.

However, if the choice is made to exceed these limits, Johns-Manville will not be responsible for any failure or damage that may occur to the roof membrane, insulation, vapor retarder, or deck, due to any deviation from these published recommendations.

Insulation loose from the deck or non-uniformly adhered not only adds to blow-off potential but does in many instances contribute to premature failure of the overlying roofing membrane. Much of this hazard can often be overcome or minimized by the use of mechanical fasteners in attaching insulation together with the use of the J-M ASBESTOGARD Vapor Retardant System. To further reduce the blow-off potential it is required that a 4'-0" min. width of insulation around the entire roof perimeter be mechanically fastened to the steel deck, in accordance with instructions contained in Factory Mutual Data 1-28. In many instances this perimeter width should be increased if abnormal wind conditions are anticipated.

Concrete decks

Poured structural concrete decks of good mix and properly placed provide a satisfactory deck for application of a roof system. Control joints properly placed, and a relatively smooth surface are the primary considerations. Concrete decks must be dry and primed with concrete primer and primer allowed to dry thoroughly before asphalt is applied.*

Pre-cast concrete slabs. (not pre-stressed) Units are

furnished that are both nailable and non-nailable. If units are out of level install a levelling fill (properly vented) to provide a satisfactory surface for the roof system. If roof membrane is to be mopped in place all joints between units should be pointed with mortar or with Industrial Roof Cement except when tongue and groove metal-bound units are used. These decks must be primed with concrete primer and primer allowed to dry thoroughly before asphalt is applied.

Pre-stressed pre-cast T or TT long span units.

These units have considerable variation in camber and in linear accuracy. Permanent deflection of this type can be progressive with applied loads such as air-conditioning equipment or other roof loads and even from the weight of pools of standing water formed in the deflected areas.

To provide a suitable surface for application of the roofing either $3/4''$ Fesco (min.) if variation in level between adjacent units is $1/4''$ or less, or a levelling fill (properly vented) must be used. If the fill is of lightweight concrete it should have a proper depth as recommended by the manufacturer of the fill and of sufficient density to receive and retain fasteners used to secure the Ventsulation Base Felt.

Lightweight Insulating Concrete Decks Poured-in Place.

The designer is urged to select his preferred type, consult the publications of the Perlite Institute, Vermiculite Institute, or the company furnishing the aggregate, write a tight specification, require application by a contractor approved or certified by the company furnishing the aggregate, and require control sampling and testing during the entire application. Some contractors offer this testing and certification in writing that the deck has been properly applied.

The relatively high casting moisture of these decks can often result in blistering and damage to the membrane. Underside venting of the lightweight deck may be recommended, however, regardless of whether a vented or non-vented lightweight deck or fill is used, Johns-Manville Ventsulation system, which provides adequate venting from the top of the deck must be employed. See specification VS-1 for proper venting details.

Lightweight aggregate concrete manufacturers, to exert more rigid and realistic controls, specify fills by density and compressive strength limits rather than by mix, which was not indicative of the true characteristics of the deck.

*See "Test for Dryness"— Page 22

The following table indicates the relationship of Density & Strength to mix:

Dry Density-PCF		Compressive Strength-PSI		Basic Mix
Perlite	Vermiculite	Perlite	Vermiculite	
33.5-40	31-37	350-500	300-500	1:4
29-32	-	234-340	-	1:5
24-28	22-28	140-200	125-225	1:6

Any deck or fill with a density of 32 PCF or greater is considered to be nailable. There are several types of expanding fasteners available, for securement of Ventsulation Felt to these. Fasteners meeting deck manufacturers criteria are Es-Products, Zono-Tite (MK-111) and E. G. Insuldeck Loc-Nails. To meet Roofing Industry Standards the withdrawal resistance of fasteners must provide a 40 lb. or more per square foot resistance to uplift of the roofing membrane. If this requirement cannot be assured then field tests on the deck in question will have to determine the performance of the fastening system.

If the deck or fill density or fastener retention are up to approved standards any nailable Built-Up roofing specification must be used, if not, Johns-Manville will not accept the substrate as an acceptable base over which to install a Built-Up roof.

To minimize the possibility of blistering and moisture failure of the roof, precautions must be taken to have the deck well cured and dry before application of the roofing. Both the Perlite & Vermiculite Institutes recommend a 3-5 day drying period of suitable weather.

Test for Dryness: Should the Architect or Owner desire a basis upon which to determine the dryness of a deck, poured or precast, the following test procedure is reproduced as a guide:

(1) Foaming: When poured on the surface to which felts are to be applied, the bitumen, heated to 350 to 400 degrees F., shall not foam upon contact with the surface.

(2) Strippability: After the bitumen used in the foaming-test application has cooled to ambient temperature, the coating shall be tested for adherence. Should any portion of the sample be readily stripped clean from the deck or insulation, the surface shall not be considered dry and application shall not be started. Should rain occur during application, the work shall be stopped and shall not be resumed until the deck has been retested by the methods specified above and found to be dry.

Cellular Concrete.

This type of deck is a lightweight, insulating concrete which is foamed during installation and entrapped air cells give it its insulating characteristics. To be a satisfactory substrate for a Built-Up roof it must have sufficient density to receive and retain nails securing the Ventsulation Felt, which is the only acceptable base felt for use over this type of deck.

Deck manufacturer or applicator must furnish J-M with a warrantee assuming responsibility for performance of the deck.

Wood decks

Board Decks should be kiln dried tongue and groove lumber securely fastened to purlins for adequate strength and rigidity in the deck. Nailing must be done with long enough nails to secure the decking without working out of the wood. Preservative treatment must be of a non-oily, non-creosote type. Warped and split boards must not be used, and there must be no large knot holes. Cracks over ¼" in width, end joints ½" or more in width, and all knot holes must be covered with sheet metal nailed in place.

A dry sheathing paper must be laid over the deck.

All wood board decks shrink to some extent and expand and contract with changes in moisture content. There must be a divorcement between deck and roof membrane to accommodate this movement. If the base felt sticks to the boards the stress may be great enough to split the roof membrane. The dry paper or the heavy base felt also help prevent possible bitumen migration and drippage through the deck.

Plywood Decks provide a good surface for receiving the roof membrane. Plywood should conform to the specifications shown in Table D of the APA. It should be at least ½" in thickness. Plywood decks do not require a sheathing paper. Base felt must be nailed, or *"sprinkle-mopped" (in Region 3 only)* with at least 10 lbs. per square of steep grade asphalt.

TABLE D — Plywood Roof Decks for Guaranteed Roofs

Allowable grades	Panel Identification Index	Plywood thickness (inches) (b)	Spacing of supports (inches) (c)	Edge Support — Plyclips (number as shown), blocking, tongue & groove or other
C-C EXT-APA	24/0	3/8, 1/2	16	1
STR. I C-C EXT-APA	30/12	5/8	24	1
STR. I C-D INT-APA	32/16	1/2, 5/8	24	1
C-D INT-APA	36/16	3/4	24	1
C-D INT-APA w/ext. glue	42/20	5/8, 3/4, 7/8	32	1
C-D INT-APA w/interm. glue	48/24	3/4, 7/8	48	2

(a) Roofing nails per roofing manufacturer's recommendation. Annular ring-shank nails are recommended.
(b) Plywood continuous over two or more spans, grain of face plies across supports.

Reduced spans generally are recommended where roofing is to be guaranteed by a performance bond. Manufacturer's recommendations are shown in Table D. A ½" plywood deck nailed to joists spaced 24" o.c. with built-up roofing has been tested by Underwriters' Laboratories to a 90 psf wind uplift.

Gypsum decks

Poured Gypsum Decks. Gypsum decks are usually placed over a bulb-tee structure. The normal shrinkage occurring in the deck during the curing stage causes cracks to occur at the weakest points — generally over or adjacent to the bulb tees.

While a roof cannot be designed to accommodate all stress likely in a gypsum deck, it can be designed to accommodate normal stress set up when the deck relieves the stress through hairline breaks well distributed over the entire deck area. It is hazardous to cement a roof solidly to a gypsum deck for such cementing allows the deck stress to be transmitted directly to the roofing membrane. A nailed specification must be used to permit deck stress to be distributed over as wide an area of roofing membrane as possible and thus minimize the possibility of roof breaks.

Poured gypsum decks do not attain maximum strength for nail retention for several weeks after application. To prevent blow-off and nail back-out during these early stages it is important to use a fastener which develops not less than 40 lbs. holding power initially. Present field experience indicates that the 1½" ES Nail-Tite, Type A has good initial pull-out resistance. If this or any fastener is to be considered it should be checked carefully with the fastener manufacturer and the gypsum deck manufacturer for adequacy. Minimum deck thickness for this type fastener is 2".

Addition of a minimum ¾" layer of Fesco over the nailed base felt is recommended as Johns-Manville will not be responsible for leaks resulting from splits in the built-up roofing which are caused by cracking of the gypsum deck after the roofing has been applied regardless of the cause of deck cracking.

Pre-Cast Gypsum. These are factory formed units usually tongue and groove metal bound. The manufacturer's recommendations should be followed carefully in design of the supporting structure and in placement and attachment. Pre-cast gypsum provides a satisfactory base for nailing. Fasteners to be considered should be checked with the fastener manufacturer and the deck manufacturer for suitability.

Structural Wood Fiber Deck Units

This decking material is made from fibrous wood bonded together with a resinous or cementitious binder. Johns-Manville will only accept those units which are manufactured using a cement binder as a satisfactory substrate for a Built-Up Roof.

As with most wood products, it tends to vary dimensionally with moisture content. As the units are quite porous, moisture passes readily through it from the underside.

Increased usage has emphasized certain problems in application of the built-up roofing membrane.

Formation of wrinkles resulting from the accumulation of condensation in the open joints of the deck; the possibility of bitumen drippage through the joints or the main body of the deck units; and penetration of objectionable odors into the interior of the building.

To minimize these problems in most areas, and provide a satisfactory base for the roof membrane, it is required that a minimum ¾" layer of Fesco be installed over the Base Sheet.

Step-downs from slab to slab must be levelled off with a screed coat, trowelled to a featheredge finish. The screed coat shall be of a mix approved by the deck manufacturer, and shall extend out on the surface so as to provide a gradual transition between slabs, where necessary.

In Region 1 the only method of installing a built-up roof over these decks is by Mechanical Attachment as follows:

A. A base sheet shall be attached to the deck using mechanical clinching type fasteners.

B. A minimum ¾" layer of Fesco shall be mopped to the base felt.

C. Subsequent courses of the roofing membrane are to be applied immediately in strict accordance with the specification selected.

As an alternate, in Region 2 & 3 only, use any nailable roof specification.

Above-Deck Fills

Lightweight Aggregate Fills:

These fills consist primarily of Perlite or Vermiculite aggregate bonded into a workable mass with asphalt. They are used primarily over structural decks as insulation, levelling fills, and to provide drainage. They have no structural strength and should be applied only over a firm, sound structural deck. As with any lightweight, porous mix these fills absorb and hold moisture. They must be dry before the roof is applied.

Asphalt type fills must be tamped or rolled into place. The density depends greatly upon the efficiency of the compaction. Such fills will not receive and retain nails.

Cants should not be formed with Deck Fill, but should be treated wood or of preformed rigid insulation.

Gravel surfaced roofs only are approved for application over these fills. Certification must be provided to Johns-Manville by the Deck Fill Manufacturer that an approved applicator has installed the fill in accordance with their specifications.

Others: Any type of deck or fill not covered by the discussions above must be submitted to Johns-Manville prior to specification to determine its suitability to receive the roofing membrane. It should be remembered that the selection of the roof deck and its structural performance are the responsibilities of the designer of the building. J-M does not accept any responsibility for the performance of the roof deck.

Wood nailers

Wood nailers or curbs shall be installed by others at all eaves, gable ends and openings in the roof for the securing of roofing felts, edging, gravel stops, and roof fixtures. If roof insulation is involved these wood members shall be the same thickness as the insulation.

Roof edges flush with the roofing membrane are not recommended. The use of tapered edging strips is recommended to direct water away from roof edges to interior drains. Wood nailers should be equal in thickness to tapered edging strips and 2″ wider than the flange of metal edging strips or gravel stops to provide adequate nailing.

On non-nailable decks where the incline is such that nailing of roofing felts is required, 2″ and over for smooth surfaced roofs and 1″ and over for gravel surfaced roofs, wood nailing strips shall be provided at ridge and at intermediate points not exceeding 20′-0″ centers. On gravel surfaced roofs, if slope exceeds 2″ per foot it is recommended nailing strips be on 10′-0″ centers to reduce possibility of slippage. If roof insulation is involved on these inclines, nailing strips the same thickness as the insulation shall be run horizontally to receive the insulation and retain nails securing the felts. On insulated roof decks where the incline is 3″ per foot and over nailing strips shall be installed 4′-0-¼″ from inside face to inside face. Roofing Felts are installed parallel to the slope, at right angles to the wood nailers.

Nailing strips and wood edging or curbs shall be of treated wood by the pressure process with a water-borne salt as approved by the American Wood-Preserver's Assn. Oil based preservatives such as creosote are not acceptable as they are not compatible with asphalt roofing components.

Roof Deck and Fastener Data

More comprehensive data now available concerning wind action on structures & roof coverings dictates that Designers & Architects be more aware of possible damage to roofs. Factory mutual data sheets 1-7, 1-28, 1-47 & 1-49 call attention to the fact that some geographical locations are subject to wind conditions that require increased fastening means at critical areas. It is the Designer & Architects responsibility to consider these factors in his design of a roofing system.

The table on this page indicates the proper built-up roof specifications for application over various roof decks. Not every roof deck has been included by trade name. However, the unlisted ones can be readily identified in one of the major groupings of decks and treated as indicated.

Also shown is a table of fastener recommendations for securement of roofing to nailable decks. This chart identifies the fasteners referred to in the table.

As density of decks vary and in many cases more than one type of fastener is specified, field tests should be conducted to determine the most effective fastener.

Since we do not manufacture either the fastener or the deck material we can not assume any responsibility for their performance. The suggested fasteners for various uses are a guide only and are not to be considered as guaranteed methods of securement by Johns-Manville.

TYPE OF JOHNS-MANVILLE BUILT-UP ROOF		NAILABLE SPECS	NON-NAILABLE SPECS	INSULATION	SPECIAL INSTRUCTIONS	FASTENER RECOMMENDATION
Smooth Surface-Asbestos-Asphalt	½"-6" per ft.	100,102	101,103	101-I,103-I		
Smooth Surface-Asbestos-Aquadam	up to ½" per ft.	150,152	151,153	151-I,153-I		
Mineral Surface-Fiber Glass	¼"-6" per ft.	400,402	401,403	401-I,403-I		
		404,406	405,407	405-I,407-I		
		408	409	409-I		
Smooth Surface-Asbestos-Cold Appl	½"-6" per ft.	2000	2001	2001		
Gravel Surface-Asbestos-Aquadam	up to ½"	600,630	601,631	601-I,631-I		
Gravel Surface-Asbestos Asphalt	½" to 3"	3000	3001	3001-I		
Gravel Surface-Organic-Aquadam	up to ½"	800, 802	801, 803	801-I, 803-I		
Gravel Surface-Organic-Asphalt	½" to 3"	900, 902	901, 903	901-I, 903-I		
ROOF DECK						
Wood-Tongue & Groove Sheathing		•				1,3,4,7,8,11
Plywood		•				3,4,7,8,11
Gypsum-Poured (1-7 Days)		•				9
Poured (Dry)		•				6,12
Precast-Metal Edge Plank		•				4,12
Concrete-Poured			•	•		
Precast-Alaslab			•	•	See Note "A"	
Doxplank			•	•	See Note "A"	
Dulite			•	•	See Note "A"	
Flexicore			•	•	See Note "A"	
CPC Channel Slab			•	•	See Note "A"	
Span-Deck			•	•	See Note "A"	
Spancrete			•	•	See Note "A"	
Pre-Stressed Concrete "T" or "TT"			•	•	See Note "A"	
Pre-cast Lightweight Concrete						
Calsi-Crete		•				5,6,2
Castlite		•				5,6,2
Creteplank		•				5,6,2
Cantilite		•				5,6,2
Federal Featherweight		•				5,6,2
CPC Concrete Plank		•				5,6,2
Lightweight Poured Concrete						
Perlite		•			See Note "B"	6,10,2
Vermiculite		•			See Note "B"	6,10,2
Zonolite		•			See Note "B"	6,10,2
Cellular Concrete		•			See Note "B"	6,10,2
Structural Wood Fiber						
Fibroplank		•			See Note "C"	5,6,2
Petrical		•			See Note "C"	5,6,2
Permadeck		•			See Note "C"	5,6,2
Fibertex		•			See Note "C"	5,6,2
Asphalt Lightweight Aggregate						
Dri-Pac			•		See Note "D"	—
All Weathercrete			•		See Note "D"	—

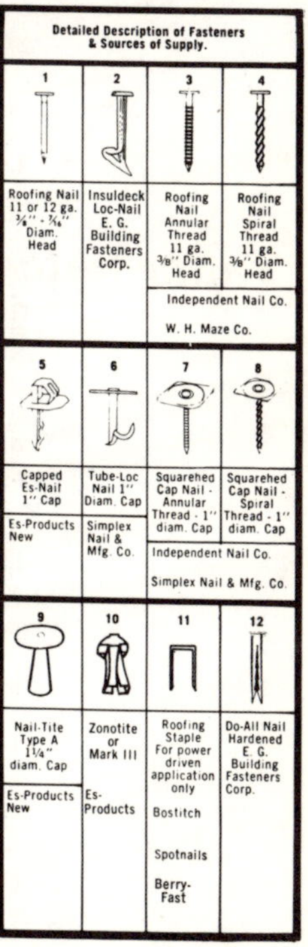

Detailed Description of Fasteners & Sources of Supply.
1
Roofing Nail 11 or 12 ga. 3/8" - 7/16" Diam. Head
2
Insuldeck Loc-Nail E. G. Building Fasteners Corp.
3
Roofing Nail Annular Thread 11 ga. 3/8" Diam. Head
4
Roofing Nail Spiral Thread 11 ga. 3/8" Diam. Head
Independent Nail Co.
W. H. Maze Co.
5
Capped Es-Nail 1" Cap
Es-Products New
6
Tube-Loc Nail 1" Diam. Cap
Simplex Nail & Mfg. Co.
7
Squarehed Cap Nail - Annular Thread - 1" diam. Cap
8
Squarehed Cap Nail - Spiral Thread - 1" diam. Cap
Independent Nail Co.
Simplex Nail & Mfg. Co.
9
Nail-Tite Type A 1¼" diam. Cap
Es-Products New
10
Zonotite or Mark III
Es-Products
11
Roofing Staple For power driven application only
Bostitch
Spotnails
Berry-Fast
12
Do-All Nail Hardened E. G. Building Fasteners Corp.

Note "A" A Layer of FescoBoard or a Leveling Fill of Lightweight Concrete may be required prior to application of roofing.

Note "B" Over these Fills Ventsulation Felt must be used as the base felt.

Note "C" A Base Sheet must first be applied over these units & secured by mechanical fasteners as recommended above. See page 26 for complete roofing instructions.

Note "D" Limited to gravel surface specifications only.

Expansion Joints

While every designer of a building considers the probable thermal movement of the structure in his design and compensates for it by placing construction and expansion joints in the structure, he often fails to consider whether or not the allowable limit in the steel or masonry exceeds the limit of the roofing membrane. While it is difficult if not impossible to establish a limit for waterproofing membranes experience has indicated general rules to follow:

1. At each construction joint or expansion joint in masonry or steel the roof should have a corresponding expansion joint.

2. If the length or width of a section of unbroken roof area exceeds 200 feet an expansion joint should be installed near the center of each direction at points not over 200 feet apart.

3. In the case of el, tee, or similar shaped roof decks an expansion joint should be provided at each intersection where the roof substrate changes direction.

4. At the intersection between different types of roof deck materials and between existing roofs and new additions.

Vapor Retarders

The decision to use or not to use a vapor retarder is the responsibility of the designer. Subsequent choice of the type of vapor retarder is part of that responsibility. While the vapor retarder is not part of the roof membrane, its absence or its performance may under certain conditions results in damage to the roof membrane as well as to the underlying roof insulation.

Vapor retarders are normally considered to be necessary for any building in any area where the January average mean temperature is 40°F. or below, and when the building is designed for human occupancy. However, in any area, a building containing high moisture producing processes or a building in which a high relative humidity is maintained is generally considered to require a vapor retarder.

If a vapor retarder is specified it should be a good retarder providing a permeability approaching zero as closely as possible. A minimum retarder would consist of a #15 Asphalt Saturated Felt uniformly mopped to the deck with hot 190 or 220 asphalt at a rate of 23 lb. per square and then top coated with an additional mopping of hot asphalt at the same rate. High humidity conditions may require a better retarder. Prepared vapor retarders must have a perm-rating of no higher than 0.5 to be significant.

Where a fire rated vapor retarder system is required to qualify for Factory Mutual Class I Construction over metal decks the use of Johns-Manville ASBESTOGARD Vapor Retardant System is mandatory.

A retarder must be of such character that the insulation can be securely attached to it by practical methods. In some instances mechanical fasteners may be used to fasten insulation through the retarder into the deck.

Roof Insulation

The basic relationship of insulation to the roof membrane is that it forms the sub-strate for the membrane. It must therefore provide a surface free of wide joints and breaks, and must be of such surface character that the membrane may be firmly attached by some practical method.

It is the responsibility of the designer to design and specify the insulation layer and the vapor retarder system (if any) to meet the expected building occupancy requirements and to conform to structural requirements such as spanning deck openings, etc. Poor wind uplift resistance from improper or inadequate installation can result in actual blow-off or serious damage to the roof membrane. If the designer expects unusual occupancy conditions or unusual exposure to wind forces, he should consult the insulation manufacturer and the built-up roofing manufacturer prior to writing his specifications. Roof blow-off or damage due to poorly or improperly secured insulation is not covered by a roofing guarantee.

Acceptance in no way implies a guarantee upon performance of the insulation or assumption of any responsibility for damage to the insulation or damage to the roof membrane due to inadequate performance or poor installation of the insulation.

Cold Weather Precautions

Roof application below 45°F. can result in problems. Special measures must be taken to insure proper performance of the roofing system. Any moisture that could cause poor adhesion, skips in moppings, or entrapment within the system must be removed from the substrate. Bitumens tend to chill quickly on a cold deck, so components of the roofing system must be installed rapidly, close to the mop, and well embedded and broomed. Bitumen must not be over-heated to compensate for rapid cooling and hoses and buggies must be insulated. The "windchill" factor must also be considered as this can also cause temperature drop.

Roofing Specifications

Suggested Form of Specification	33
Roof Finder Index	34
Underwriters' Laboratories Inc. Classifications	36
Smooth Surface Roofs	41
Mineral Surface Roofs	79
Gravel Surface Roof	113
Ventsulation Felt	157
Flashings	163
Re-Roofing	181
Temporary Roofing	191
Asbestogard	193
Waterproofing & Damp Proofing	197
Roof Insulation	203

Suggested form of specification

Roofing

Before roofing application is begun, the roofing contractor shall inspect the roof deck carefully. It shall be firm, dry, free of foreign material which would interfere with the roofing application, and reasonably smooth. All cracks, breaks, holes, or other unusual irregularities in the surface shall be reported to the General Contractor for remedy before roofing work is begun. Any disagreement on the condition of the deck shall be referred to the Architect for decision. Installation of cants, metal fittings, and similar work affecting the roofing shall be complete before work begins.

Roofing and Flashing Materials shall be applied in accordance with the following Johns-Manville specifications:

    Roof   J-M Spec No. _____
    Flashings   J-M Spec _____

Alternates proposed as equals to the above specifications shall be submitted in writing to the Architect in full detail together with explanation and reasons for such changes. Alternates will not be acceptable until approved in writing by the Architect.

No deviation from the approved specifications as written in the manufacturer's current roofing specifications manual may be made without prior approval of the architect, and in the event a guarantee is required, of the manufacturer in writing.

If a Roof Guarantee is required add the following:

Upon completion of the roofing the roofer shall furnish the manufacturer's guarantee for _____ years from date of completion of the roof, with a maximum liability of $_____per square.

Roof Finder Index

Roof Slope	Deck	Eligible for a Guarantee-Years*	For Use in Region	Spec. No.	Construction	Total Max. Wt. Per Square	Page
	SMOOTH SURFACE — ASBESTOS						
½" To 6"	Nailable	20 15	All	100 102	1 Asbestos Base, 3 Finishing, Asphalt 1 Asbestos Base, 2 Finishing, Asphalt	182 149	45 51
	Non-Nailable	20 15	All	101 103	1 Asbestos Base, 2 Finishing, Asphalt 1 Planet Base, 2 Finishing, Asphalt	167 154	47 53
	Insulation	20 15	All	101-I 103-I	1 Asbestos Base, 2 Finishing, Asphalt 1 Planet Base, 2 Finishing, Asphalt	177 187	49 55
Up To ½"	Nailable	20 15	All	150 152	1 Asbestos Base, 3 Finishing, Aquadam 1 Asbestos Base, 2 Finishing, Aquadam	182 149	57 67
	Non-Nailable	20 20 15	All 3 All	151 154 153	1 Asbestos Base, 3 Finishing, Aquadam 1 Special Ctd. Base, 2 Finishing, Aquadam 1 Asbestos Base, 2 Finishing, Aquadam	205 164 154	59 63 69
	Insulation	20 20 15 15	All 3 All 3	151-I 155-I 153-I 156-I	1 Asbestos Base, 3 Finishing, Aquadam 4 Finishing, Aquadam 1 Asbestos Base, 2 Finishing, Aquadam 3 Finishing, Aquadam	205 170 167 132	61 65 71 73
	SMOOTH SURFACE — ASBESTOS — COLD APPLICATION						
½" To 6"	Nailable		All	2000	1 Planet Base and 2 C. A. Asbestos Cold Application Cement	150	75
	Non-Nailable & Insulation		All	2001	1 Planet Base and 2 C. A. Asbestos Cold Application Cement	169	77
	GRAVEL SURFACE — ORGANIC						
Up To ½"	Nailable	20 15	All	800 802	1 Planet, 3 Organic, Aquadam 1 Planet, 2 Organic, Aquadam	617 579	133 139
	Non-Nailable	20 15	All	801 803	1 Planet, 3 Organic, Aquadam 1 Planet, 2 Organic, Aquadam	640 602	135 141
	Insulation	20 15	All	801-I 803-I	1 Planet, 3 Organic, Aquadam 1 Planet, 2 Organic, Aquadam	650 602	137 143
½" To 3"	Nailable	20 15	All	900 902	1 Planet, 3 Organic, Asphalt 1 Planet, 2 Organic, Asphalt	617 579	145 151
	Non-Nailable	20 15	All	901 903	1 Planet, 3 Organic, Asphalt 1 Planet, 2 Organic, Asphalt	640 602	147 153
	Insulation	20 15	All	901-I 903-I	1 Planet, 3 Organic, Asphalt 1 Planet, 2 Organic, Asphalt	650 602	149 155

These specifications are eligible for a Guarantee only when, in the opinion of an authorized J-M Representative, all conditions listed in "General Instructions" of this Specification Manual have been met.

Roof Slope	Deck	Eligible for a Guarantee-Years*	For Use in Region	Spec. No.	Construction	Total Max. Wt. Per Square	Page
	**MINERAL SURFACE**						
¼" To 6"	Nailable	20	All	400	1 Asbestos Base, 2 Finishing, 1 Mineral	219	81
		15	All	402	1 Planet, 2 Organic, 1 Mineral	219	87
		20	3	404	1 Asbestos Base, 1 Finishing, 1 Mineral	191	93
		20	3	406	2 Finishing, 1 Mineral	163	99
		15	3	408	2 Organic, 1 Mineral	163	105
	Non-Nailable	20	All	401	1 Asbestos Base, 2 Finishing, 1 Mineral	242	83
		15	All	403	1 Planet, 2 Organic, 1 Mineral	250	89
		20	3	405	1 Asbestos Base, 1 Finishing, 1 Mineral	191	95
		20	3	407	2 Finishing, 1 Mineral	163	101
		15	3	409	1 Planet, 1 Organic, 1 Mineral	191	107
	Insulation	20	All	401-I	1 Asbestos Base, 2 Finishing, 1 Mineral	252	85
		15	All	403-I	1 Planet, 2 Organic, 1 Mineral	252	91
		20	3	405-I	1 Asbestos Base, 1 Finishing, 1 Mineral	214	97
		20	3	407-I	2 Asbestos, 1 Mineral	189	103
		15	3	409-I	2 Organic, 1 Mineral	189	109
¼" To 6"	Any Acceptable Deck	20	3	420 VS1	1 Ventsulation, 1 Finishing, 1 Mineral	223	111
	**GRAVEL SURFACE — ASBESTOS**						
Up To ½"	Nailable	20	2, 3	600	1 Asbestos Base, 2 Finishing, Aquadam	589	115
		15	1				
		20	All	630	1 Asbestos Base, 3 Finishing, Aquadam	627	121
	Non-Nailable	20	All	601	1 Asbestos Base, 2 Finishing, Aquadam	597	117
		20	All	631	1 Asbestos Base, 3 Finishing, Aquadam	645	123
	Insulation	20	All	601-I	1 Asbestos Base, 2 Finishing, Aquadam	612	119
		20		631-I	1 Asbestos Base, 3 Finishing, Aquadam	658	125
½" To 3"	Nailable	20	2, 3	3000	1 Asbestos Base, 2 Finishing, Asphalt	589	127
		15	1				
	Non-Nailable	20	All	3001	1 Asbestos Base, 2 Finishing, Asphalt	597	129
	Insulation	20	All	3001-I	1 Asbestos Base, 2 Finishing, Asphalt	612	131

These specifications are eligible for a Guarantee only when, in the opinion of an authorized J-M Representative, all conditions listed in "General Instructions" of this Specification Manual have been met.

Johns-Manville
JM

**Underwriters'
Laboratories, Inc.
Classifications**

Built-Up Roofs

Underwriters' Laboratories Inc. Classification

Johns-Manville Built-Up Roofs

Class A

J-M Roof Specification Number	Maximum Roof Slope						Special Requirements
	Non-Combustible Deck			Combustible Deck			
	Uninsulated	¾" Min. Fesco	Fesco-Foam C.15 to C.05	Uninsulated	¾" Min. Fesco	Fesco-Foam C.15 to C.05	
Mineral Surface — GlasKap							
400, 401, 404, 405, 406, 407, 420-VS1	1"	NA	NA	1"	NA	NA	
401-I, 405-I	NA	1"	1"	NA	1"	1"	
407-I	NA	1"	1"	NA	NA	NA	
403-I, 409-I	NA	1"	1"	NA	NA	NA	
402, 408, 403, 409	1"	NA	NA	NA	NA	NA	
Mineral Surface — Flexstone							
400, 401, 404*, 405*	1½"	NA	NA	1½"	NA	NA	*Use Coated Asbestos Base Felt Only over combustible deck
406, 407, 420-VS1	1½"	NA	NA	NA	NA	NA	
401-I, 405-I*	NA	1½"	1½"	NA	1½"	1½"	
407-I	NA	1½"	1½"	NA	NA	NA	
403-I, 409-I	NA	1½"	1½"	NA	NA	NA	
402, 408, 403, 409	1½"	NA	NA	NA	NA	NA	
Gravel Surface — Asbestos							
600, 601, 630, 631	½"	NA	NA	½"	NA	NA	
601-I, 631-I	NA	½"	½"	NA	½"	½"	
3000, 3001	3"	NA	NA	3"	NA	NA	
3001-I	NA	3"	3"	NA	3"	3"	
Gravel Surface — Organic							
800, 801, 802, 803	½"	NA	NA	½"	NA	NA	
801-I, 803-I	NA	½"	½"	NA	½"	½"	
900, 901, 902, 903	3"	NA	NA	3"	NA	NA	
901-I, 903-I	NA	3"	3"	NA	3"	3"	
Smooth Surface Asbestos — Hot Mopped Coating							
100, 101, 102, 103	1"	NA	NA	NA	NA	NA	For Top Coating use 20 lbs. of same melting point Asphalt as for interply mopping.
101-I, 103-I	NA	1"	½"	NA	NA	NA	
150, 151, 152, 153, 154	½"	NA	NA	NA	NA	NA	
151-I, 153-I, 154-I, 155-I	NA	½"	½"	NA	NA	NA	
Smooth Surface Asbestos — Emulsion Coating — Topgard B							
100	2"	NA	NA	NA	NA	NA	Top Coating-Fibrated Topgard 2 to 3 Gals. per square.
101, 102, 103	3"	NA	NA	NA	NA	NA	
101-I, 103-I	NA	2"	2"	NA	NA	NA	
Smooth Surface Asbestos — Fibrated Coating — Topgard Type F or Aluminum Roof Coating							
100, 101, 102, 103	2"	NA	NA	NA	NA	NA	Top Coating-Fibrated Topgard 1 to 1½ Gals. per square — OR Fibrated Aluminum Roof Coating ¾ to 1¼ Gals. per square.
101-I, 103-I	NA	2"	2"	NA	NA	NA	
150, 151, 152, 153, 154	½"	NA	NA	NA	NA	NA	
151-I, 153-I, 154-I, 155-I	NA	½"	½"	NA	NA	NA	
Smooth Surface Asbestos — Cutback Asphalt Coating — Topgard Type C							
100, 101, 102, 103	1½"	NA	NA	NA	NA	NA	Top Coating-Topgard Type C, 1 Gal. per square
101-I, 103-I	NA	1½"	1"	NA	NA	NA	

NA = Not Applicable

Notes

1. J-M AQUADAM, 170, 190 and 220 asphalt may be used interchangeably as required. However, AQUADAM is limited to ½-in. slopes.

2. On non-combustible decks, FESCO, FESCO/Foam, and other UL labeled and listed roof insulations may be used.

3. On combustible decks, only FESCO or FESCO/Foam may be used. Other insulations are not rated.

4. Base sheets may be solid, sprinkle or spot mopped or attached by mechanical fasteners as dictated by good roofing practice.

5. A non-combustible deck is metal, concrete or poured gypsum.

6. Vapor retardant systems including Asbestogard, organic felt, asbestos felt, etc. do not affect the UL listing.

7. For fire protection purposes, dry or slip sheets of kraft paper over wood decks are considered to be insignificant and do not affect UL listings.

8. If a roof construction is to qualify for a UL listing, all roofing materials must bear UL labels. Notification of this requirement must be made known to plants at time of ordering. UL label service is also available on bulk shipments of bitumens from refineries.

9. The information contained in this folder was accurate at the time of printing. You may also wish to refer to the *Building Materials Directory* published by Underwriters' Laboratories.

Class C

J-M Roof Specification Number	Maximum Roof Slope						Special Requirements
	Non-Combustible Deck			Combustible Deck			
	Uninsulated	¾" Min. Fesco	Fesco-Foam C.15 to C.05	Uninsulated	¾" Min. Fesco	Fesco-Foam C.15 to C.05	
Mineral Surface — GlasKap							
400, 401, 404, 405, 406, 407, 420-VS1	1½"	NA	NA	1½"	NA	NA	
401-I, 405-I	NA	1½"	1½"	NA	1½"	1½"	
407-I	NA	1½"	1½"	NA	1½"	1½"	
403-I, 409-I	NA	1½"	1½"	NA	NA	NA	
402, 408, 403, 409	1½"	NA	NA	½"	NA	NA	
Mineral Surface — Flexstone							
400, 401, 404, 405 406, 407, 420-VSI	6"	NA	NA	6"	NA	NA	
401-I, 405-I*	NA	6"	6"	NA	6"	6"	
407-I	NA	6"	6"	NA	6"	6"	
403-I, 409-I	NA	6"	6"	NA	6"	6"	
402, 408, 403, 409	6"	NA	NA	6"	NA	NA	
Gravel Surface — Asbestos							
600, 601, 630, 631	½"	NA	NA	½"	NA	NA	
601-I, 631-I	NA	½"	½"	NA	½"	½"	
3000, 3001	3"	NA	NA	3"	NA	NA	
3001-I	NA	3"	3"	NA	3"	3"	
Gravel Surface — Organic							
800, 801, 802, 803	½"	NA	NA	½"	NA	NA	
801-I, 803-I	NA	½"	½"	NA	½"	½"	
900, 901, 902, 903	3"	NA	NA	3"	NA	NA	
901-I, 903-I	NA	3"	3"	NA	3"	3"	
Smooth Surface Asbestos — Hot Mopped Coating							
100, 101, 102, 103	1"	NA	NA	1"	NA	NA	For Top Coating use 20 lbs. of same melting point Asphalt as for interply mopping.
101-I, 103-I	NA	1"	½"	NA	1"	½"	
150, 151, 152, 153, 154	½"	NA	NA	½"	NA	NA	
151-I, 153-I, 154-I, 155-I	NA	½"	½"	NA	½"	½"	
Smooth Surface Asbestos — Emulsion Coating — Topgard B							
100	6"	NA	NA	6"	NA	NA	Top Coating-Fibrated Topgard 2 to 3 Gals. per square.
101, 102, 103	6"	NA	NA	6"	NA	NA	
101-I, 103-I	NA	6"	6"	NA	6"	6"	
Smooth Surface Asbestos — Fibrated Coating — Topgard Type F or Aluminum Roof Coating							
100, 101, 102, 103	4"	NA	NA	4"	NA	NA	Top Coating-Fibrated Topgard 1 to 1½ Gals. per square — OR Fibrated Aluminum Roof Coating ¾ to 1¼ Gals. per square.
101-I, 103-I	NA	4"	4"	NA	4"	4"	
150, 151, 152, 153, 154	½"	NA	NA	½"	NA	NA	
151-I, 153-I, 154-I, 155-I	NA	½"	½"	NA	½"	½"	
Smooth Surface Asbestos — Cutback Asphalt Coating — Topgard Type C							
100, 101, 102, 103	3"	NA	NA	3"	NA	NA	Top Coating-Topgard Type C, 1 Gal. per square
101-I, 103-I	NA	3"	2½"	NA	3"	2½"	

NA = Not Applicable

Class B

J-M Roof Specification Number	Maximum Roof Slope						Special Requirements
	Non-Combustible Deck			Combustible Deck			
	Uninsulated	¾" Min. Fesco	Fesco-Foam C.15 to C.05	Uninsulated	¾" Min. Fesco	Fesco-Foam C.15 to C.05	
Mineral Surface — GlasKap							
400, 401, 404, 405, 406, 407, 420-VS1	1½"	NA	NA	1½"	NA	NA	
401-I, 405-I	NA	1½"	1½"	NA	1½"	1½"	
407-I	NA	1½"	1½"	NA	1½"	1½"	
403-I, 409-I	NA	1½"	1½"	NA	NA	NA	
402, 408, 403, 409	1½"	NA	NA	½"	NA	NA	
Mineral Surface — Flexstone							
400, 401, 404*, 405*	3"	NA	NA	3"	NA	NA	*Use Coated Asbestos Base Felt Only over combustible deck
406, 407, 420-VS1	3"	NA	NA	NA	NA	NA	
401-I, 405-I*	NA	3"	3"	NA	3"	3"	
407-I	NA	3"	3"	NA	3"	NA	
403-I, 409-I	NA	3"	3"	NA	2½"	NA	
402, 408, 403, 409	3"	NA	NA	2½"	NA	NA	
Gravel Surface — Asbestos							
600, 601, 630, 631	½"	NA	NA	½"	NA	NA	
601-I, 631-I	NA	½"	½"	NA	½"	½"	
3000, 3001	3"	NA	NA	3"	NA	NA	
3001-I	NA	3"	3"	NA	3"	3"	
Gravel Surface — Organic							
800, 801, 802, 803	½"	NA	NA	½"	NA	NA	
801-I, 803-I	NA	½"	½"	NA	½"	½"	
900, 901, 902, 903	3"	NA	NA	3"	NA	NA	
901-I, 903-I	NA	3"	3"	NA	3"	3"	
Smooth Surface Asbestos — Hot Mopped Coating							
100, 101, 102, 103	1"	NA	NA	1"	NA	NA	For Top Coating use 20 lbs. of same melting point Asphalt as for interply mopping.
101-I, 103-I	NA	1"	½"	NA	1"	½"	
150, 151, 152, 153, 154	½"	NA	NA	½"	NA	NA	
151-I, 153-I, 154-I, 155-I	NA	½"	½"	NA	½"	½"	
Smooth Surface Asbestos — Emulsion Coating — Topgard B							
100	4"	NA	NA	3"	NA	NA	Top Coating-Fibrated Topgard 2 to 3 Gals. per square.
101, 102, 103	4"	NA	NA	3"	NA	NA	
101-I, 103-I	NA	3"	3"	NA	2"	2"	
Smooth Surface Asbestos — Fibrated Coating — Topgard Type F or Aluminum Roof Coating							
100, 101, 102, 103	3"	NA	NA	3"	NA	NA	Top Coating-Fibrated Topgard 1 to 1½ Gals. per square — OR Fibrated Aluminum Roof Coating ¾ to 1¼ Gals. per square.
101-I, 103-I	NA	3"	3"	NA	2"	2"	
150, 151, 152, 153, 154	½"	NA	NA	½"	NA	NA	
151-I, 153-I, 154-I, 155-I	NA	½"	½"	NA	½"	½"	
Smooth Surface Asbestos — Cutback Asphalt Coating — Topgard Type C							
100, 101, 102, 103	2"	NA	NA	2"	NA	NA	Top Coating-Topgard Type C, 1 Gal. per square
101-I, 103-I	NA	2"	1½"	NA	2"	1½"	

NA = Not Applicable

Johns-Manville

P.O. Box 5108
Denver, Colorado 80217

Johns-Manville
Smooth-Surface Asbestos BUILT-UP ROOFS

SMOOTH-SURFACE ASBESTOS

defy the sun

defy the weather

specify the permanence of asbestos... enduring... dependable... economical

Johns-Manville

Smooth-Surface Asbestos BUILT-UP ROOFS

... there is a specification
- *to suit your particular building*
- *to withstand the weather conditions in your locality*
- *to fit your budget requirements*

Smooth Surface Asbestos Roofs are actually flexible coverings of stone ... enduring and dependable ... because they are based on that ageless mineral, asbestos.

Asbestos is a fibrous stone. It is possible to tease the fibers apart and each individual fiber is a slender rod of stone. These unique fibers, strong yet flexible, act as re-inforcing rods in the roofing felt. The result is a felt that is weatherproof, won't rot, won't dry out

Asbestos felts, as produced by Johns-Manville, contain a minimum of 85% inorganic materials. The high content of stone-like fiber means no slag or gravel is needed ... the roof can be smooth surfaced. The superficial surfacing of stone is eliminated because the smooth-surface asbestos roof has stone-like protection all-the-way-through.

Every bituminous built-up roof contains two fundamental materials ... felts and bitumens. When assembled they form the roofing membrane. The bitumens provide waterproofness to the membrane. The felts stabilize, reinforce and strengthen the membrane. Without felts the bitumen would flow, crack and generally deteriorate within a relatively short period of time. When ordinary organic felts are used, the roof is finished with a gravel surfacing to protect elements of the membrane. However, when asbestos felts are used no gravel surfacing is needed since the asbestos provides the necessary protection.

Felts consist of fibers and binders which are made into a dry felt. Then these felts are impregnated with weatherproofing bitumens.

Ordinary felts are made of organic fibers. They are hollow like tubes. All of these tiny tubes act as countless "wicks" through which the weatherproofing bitumens can be drawn off by the sun. This capillary action eventually evaporates not only the bitumen within the felt but also the bitumens between the layers of felts thus leading to failure of the roof. Also, lacking the permanence of asbestos, the organic fibers rot away when exposed.

By contrast, each fiber of asbestos is a solid rod that is immune to the sun. There is no capillary action within an asbestos roof. The stone-like fibers won't rot or decay.

A magnified section of a Smooth-Surface Asbestos Roof shows a network of fibers in each layer. These strong yet flexible fibers reinforce the roof and, most important, they protect the bitumen impregnated within the felt and also protect the layers of bitumen between the layers of felt. To prove the effectiveness of asbestos fibers, try the "scrape" test. Wherever the smooth gray surface is scraped, rich black bitumen is found underneath. And, this test can be repeated a dozen times in the same spot with the same results.

Johns-Manville has been producing roofing felts made of asbestos for over 100 years! On the record, Johns-Manville can cite many asbestos roofs that, today, are still performing satisfactorily after more than forty years of exposure to heat, cold, water and air. And, all of this service with minimum maintenance and repair costs. When evaluating true satisfaction and economy, consider the advantages of a Smooth Surface Asbestos Built-up Roof.

Johns-Manville *Smooth-Surface Asbestos Roofs offer these valuable advantages*

LIGHTWEIGHT — less than ⅓ the weight of a gravel surfaced roof.

EASY TO INSPECT — the surface is exposed. Damage is easy to find by visual inspection.

GREATER ECONOMY — with no gravel surfacing, your entire investment is for actual waterproofing materials.

BETTER WORKMANSHIP — since it is smooth surfaced, nothing is covered up. This encourages good workmanship.

EASY TO REPAIR — if damaged, a patch can be made immediately and quickly.

EASY TO MAINTAIN — the smooth surface does not catch and hold dirt or ordinary industrial waste. Any such accumulations that do occur can be easily disposed of, usually by flushing with water.

EASIER RE-ROOFING — when a new roof is desired it usually can be applied without removing the old membrane. However, if removal is decided on, there is less material to take off and dispose of.

BETTER WEATHER PROTECTION — asbestos does not rot or decay.

LONGER LIFE — the sun doesn't "dry-out" an asbestos roof.

for Industrial Buildings

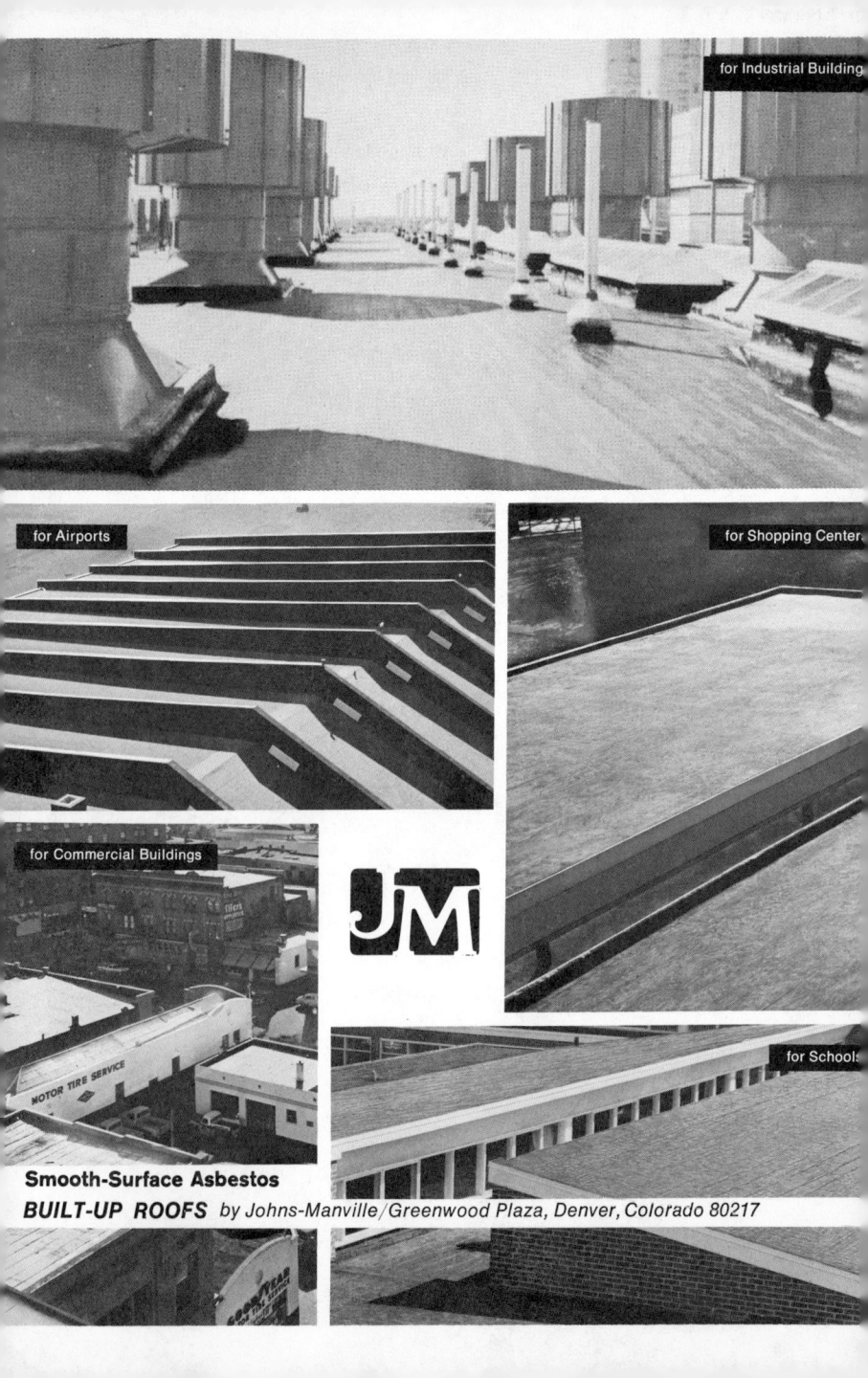

for Airports

for Shopping Centers

for Commercial Buildings

for Schools

JM

Smooth-Surface Asbestos
BUILT-UP ROOFS by Johns-Manville/Greenwood Plaza, Denver, Colorado 80217

Specification **No. 100** *for use over*

PLYWOOD OR OTHER NAILABLE DECKS

on inclines of 1/2" to 6" per foot

Johns-Manville

Smooth-Surface Asbestos BUILT-UP ROOFS

This specification is to be used over any type of structural deck (without insulation) which can receive and adequately retain nails or other types of mechanical fasteners as may be recommended by the deck manufacturer. Examples of such decks are wood, plywood, and some lightweight aggregate concrete decks* no lighter than 32 PCF density, or that provide proper nail retention.

*Ventsulation Felt must be the base felt over these decks.

Note: All information contained in "General Instructions" in the current Specification Manual for Johns-Manville Built-Up Roofs shall be considered part of this specification.

Flashings
See the section titled FLASHINGS, in the Specification Manual for J-M Built-Up Roofs.

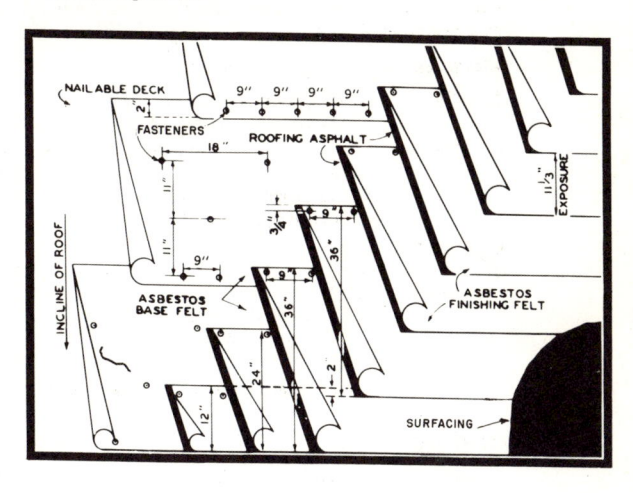

This specification is eligible for a 20-Year Guarantee only when, in the opinion of an authorized J-M Representative, all conditions listed in "General Instructions" of this Specification Manual have been met.

Application of Roofing

Over wood board decks one ply of sheathing paper must be used under the base felt next to the deck.

Regions 1 & 2: Use J-M Centurian as the base felt.

Region 3: Use either J-M Centurian or Coated Asbestos as the base felt.

First: Start at the low edge and working up the slope and perpendicular to the slope and lapping each felt 2" over the preceding one. Nail the laps at 9" centers and down the longitudinal center of each felt nail two rows of nails with rows spaced approximately 11" apart and nails staggered on approximately 18" centers. Use nails for fasteners appropriate to the type of deck.

Note: In Region 3 only if deck is plywood, base felt may be sprinkle mopped using 10 lbs. of asphalt per square.

Second: Starting at the low edge apply one 12" wide, then over that one 24" wide, then over both a full 36" wide J-M Asbestos Finishing Felt. Following felts are to be applied full width overlapping the preceding felt by 24⅔" in such manner that at least 3 plies of felt cover the base felt at any point. On slopes 2" per foot or greater nail each felt at approximately 9" centers adjacent to the back edge. Broom each felt so that it shall be firmly and uniformly set without voids into hot asphalt applied just before the felt at a minimum rate of 23 lbs per square uniformly over the entire surface.

Third: Finish the entire surface with a uniform coating of one of the surfacing options below.

Materials per 100 sq ft of roof area

FELTS:	J-M Centurian or Coated Asbestos Felt	1 layer
	J-M Asbestos Finishing Felt	3 layers
BITUMEN:	J-M 190 Asphalt (Slopes under 3" per foot)	69 lbs
	J-M 220 Asphalt (Slopes 3" or greater)	69 lbs
	Sprinkle mopping to plywood	10 lbs
SURFACING OPTIONS: (Select One)	J-M Topgard Type B — Fibrated	2-3 gals
	J-M Topgard Type F	1-1½ gals
	J-M Fibrated Aluminum Roof Coating	¾-1¼ gals
	J-M Topgard Type C	1 gal
	J-M Asphalt	20 lbs

Approximate applied Weight Min: 147 lbs. Max: 182 lbs.

Nailing

All nails or other fasteners are to be driven through tin caps unless the nail or fastener has an integral flat cap no less than 1" across.

BUILT-UP ROOFS by Johns-Manville/Greenwood Plaza, Denver, Colorado 80217

Specification **No. 101** *for use over*

CONCRETE OR OTHER NON-NAILABLE DECKS

on inclines of 1/2" to 6" per foot

Johns-Manville

Smooth-Surface Asbestos BUILT-UP ROOFS

This specification is for use over any type of structural deck which is not nailable and which offers suitable surface to receive the roof. Poured and pre-cast concrete decks require priming. This specification is not to be used over lightweight insulating concrete decks either poured or pre-cast, or over fill made of lightweight insulating concrete.

Note: All information contained in "General Instructions" in the current Specification Manual for Johns-Manville Built-Up Roofs shall be considered part of this specification.

Flashings

See section on FLASHINGS, Specification Manual for J-M Built-Up Roofs.

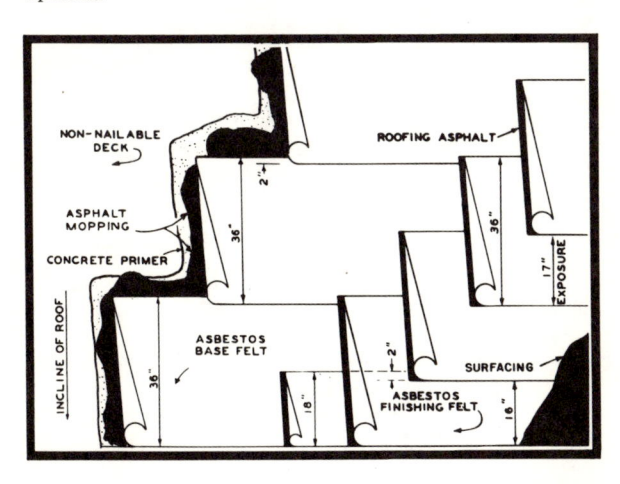

On slopes up to 2" apply finishing felts perpendicular to the slope starting at the low point of each slope. On slopes over 2" apply finishing felts parallel to the slope, nailing at the top of each run of felt on not over 9" centers. If run of felt exceeds 20' an additional line of nails shall be used at 20' intervals.

This specification is eligible for a 20-Year Guarantee only when, in the opinion of an authorized J-M Representative, all conditions listed in "General Instructions" of this Specification Manual have been met.

Application of Roofing

First: Regions 1 & 2: Use J-M Centurian as the base felt, lapping each felt 2" over the preceding one and solidly mop the full width under each ply felt with asphalt using a minimum of 23 lbs. per square.

Region 3: Use either J-M Centurian or Coated Asbestos as the base felt, lapping each felt 2" over the preceding one and spot mop the full width under each felt with asphalt using a minimum of 10 lbs. per square.

Second: Starting at the low edge (on slopes up to 2") apply one 18" wide, then over that one 36" wide J-M Asbestos Finishing Felt. Following felts are to be applied full width overlapping the preceding felt by 19" in such manner that at least 2 plies of felt cover the base felt at any point. Broom each felt so that it shall be firmly and uniformly set without voids into hot asphalt applied just before the felt at a minimum rate of 23 lbs. per square uniformly over the entire surface. On slopes over 2" per foot all felts shall be nailed at the top of each run of felt on not over 9" centers. If run of felt exceeds 20' an additional line of nails shall be used at 20' intervals.

Third: Finish the entire surface with a uniform coating of one of the surfacing options below.

Materials per 100 sq ft of roof area

CONCRETE PRIMER:	If required	1 gal
FELTS:	J-M Centurian or Coated Asbestos Base Felt	1 layer
	J-M Asbestos Finishing Felt	2 layers
BITUMEN:	J-M 190 Asphalt (Slopes under 3" per foot)	69 lbs*
	J-M 220 Asphalt (Slopes 3" or greater)	69 lbs*
SURFACING OPTIONS: (Select One)	J-M Topgard Type B — Fibrated	2-3 gals
	J-M Topgard Type F	1-1½ gals
	J-M Fibrated Aluminum Roof Coating	¾-1¼ gals
	J-M Topgard Type C	1 gal
	J-M Asphalt	20 lbs

Approximate applied Weight Min: 119 lbs. Max: 167 lbs.

*Deduct 13 lbs. if spot mopped.

Nailing

Where nailing is required, nailing strips must be provided. All nails or other fasteners are to be driven through tin caps unless the nail or fastener has an integral flat cap no less than 1" across.

BUILT-UP ROOFS by Johns-Manville/Greenwood Plaza, Denver, Colorado 80217

Regions 1, 2, & 3

Specification **No. 101-I** *for use over*

FESCO, FESCO-FOAM OR OTHER APPROVED INSULATION

on inclines of 1/2" to 6" per foot

Johns-Manville

Smooth-Surface Asbestos BUILT-UP ROOFS

This specification is for use over Fesco, Fesco-Foam or any type of approved insulation which is not nailable and which offers suitable surface to receive the roof. This specification is not to be used over light-weight aggregate concrete decks either poured or pre-cast, or over fill made of lightweight insulating concrete.

Note: All information contained in "General Instructions" in the current Specification Manual for Johns-Manville Built-Up Roofs shall be considered part of this specification.

Flashings

See section on FLASHINGS, Specification Manual for J-M Built-Up Roofs.

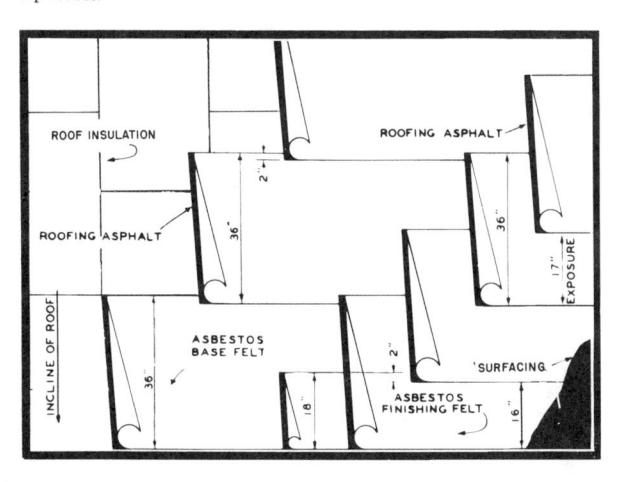

On slopes up to 2" apply felts perpendicular to the slope starting at the low point of each slope. On slopes over 2" apply felts parallel to the slope, nailing at the top of each run of felt on not over 9" centers. If run of felt exceeds 20' an additional line of nails shall be used at 20' intervals.

This specification is eligible for a 20-Year Guarantee only when, in the opinion of an authorized J-M Representative, all conditions listed in "General Instructions" of this Specification Manual have been met.

Application of Roofing

Regions 1 & 2: Use J-M Centurian as the base felt.

Region 3: Use either J-M Centurian or Coated Asbestos as the base felt.

First: Lap each felt 2" over the preceding one. Mop the full width under each felt with the appropriate hot J-M asphalt using a minimum of 33 lbs. per square.

Second: Starting at the low edge (on slopes up to 2") apply one 18" wide, then over that one 36" wide J-M Asbestos Finishing Felt. Following felts are to be applied full width overlapping the preceding felt by 19" in such manner that at least 2 plies of felt cover the base felt at any point. Broom each felt so that it shall be firmly and uniformly set without voids into hot asphalt applied just before the felt at a minimum rate of 23 lbs. per square uniformly over the entire surface. On slopes over 2" per foot all felts shall be nailed at the top of each run of felt on not over 9" centers. If run of felt exceeds 20' an additional line of nails shall be used at 20' intervals.

Third: Finish the entire surface with a uniform coating of one of the surfacing options below.

Materials per 100 sq ft of roof area

FELTS:	J-M Centurian or Coated Asbestos Base Felt	1 layer
	J-M Asbestos Finishing Felt	2 layers
BITUMEN:	J-M 190 Asphalt (Slopes under 3" per foot)	79 lbs*
	J-M 220 Asphalt (Slopes 3" or greater)	79 lbs*
SURFACING OPTIONS: (Select One)	J-M Topgard Type B — Fibrated	2-3 gals
	J-M Topgard Type F	1-1½ gals
	J-M Fibrated Aluminum Roof Coating	¾-1¼ gals
	J-M Topgard Type C	1 gal
	J-M Asphalt	20 lbs

Approximate applied Weight Min: 142 lbs. Max: 177 lbs.

*If over Fesco-Foam deduct 10 lbs.

Nailing

Where nailing is required, nailing strips must be provided. All nails or other fasteners are to be driven through tin caps unless the nail or fastener has an integral flat cap no less than 1" across.

BUILT-UP ROOFS by Johns-Manville/Greenwood Plaza, Denver, Colorado 80217

Regions 1, 2, & 3

*Specification **No. 102** for use over*

PLYWOOD OR OTHER NAILABLE DECKS

*on inclines of **1/2"** to **6"** per foot*

Johns-Manville
Smooth-Surface ASBESTOS *BUILT-UP ROOFS*

This specification is to be used over any type of structural deck (without insulation) which can receive and adequately retain nails or other types of mechanical fasteners as may be recommended by the deck manufacturer. Examples of such decks are wood, plywood, and some lightweight aggregate concrete decks* no lighter than 32 PCF density, or that provide proper nail retention.

*Ventsulation Felt must be the base felt over these decks.

Note: All information contained in "General Instructions" in the current Specification Manual for Johns-Manville Built-Up Roofs shall be considered part of this specification.

Flashings

See section on FLASHINGS, Specification Manual for J-M Built-Up Roofs.

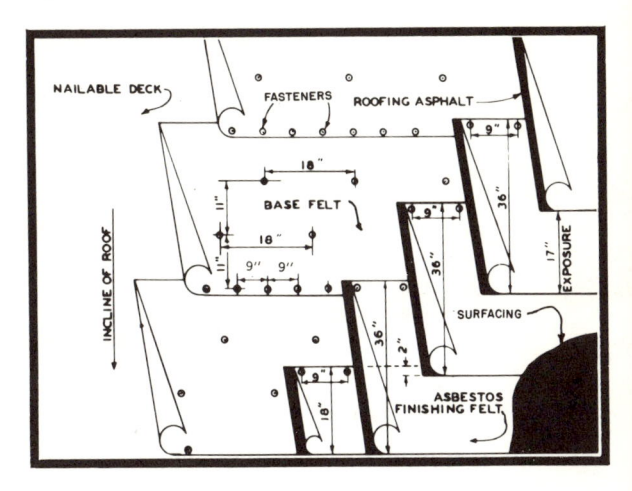

This specification is eligible for a 15-Year Guarantee only when, in the opinion of an authorized J-M Representative, all conditions listed in "General Instructions" of this Specification Manual have been met.

Application of Roofing

Over wood board decks one ply of sheathing paper must be used under the base felt next to the deck.

Regions 1 & 2: Use J-M Centurian as the base felt.

Region 3: Use either J-M Centurian or Coated Asbestos as the base felt.

First: Start at the low edge and working up the slope and perpendicular to the slope and lapping each felt 2" over the preceding one. Nail the laps at 9" centers and down the longitudinal center of each felt nail two rows of nails with rows spaced approximately 11" apart and nails staggered on approximately 18" centers. Use nails for fasteners appropriate to the type of deck.

Note: In Region 3 only if deck is plywood, base felt may be sprinkle mopped using 10 lbs. of asphalt per square.

Second: Starting at the low edge apply one 18" wide, then over that one full 36" wide J-M Asbestos Finishing Felt. Following felts are to be applied full width overlapping the preceding felt by 19" in such manner that at least 2 plies of felt cover the base felt at any point. On slopes 2" per foot or greater nail each felt at approximately 9" centers adjacent to the back edge. Broom each felt so that it shall be firmly and uniformly set without voids into hot asphalt applied just before the felt at a minimum rate of 23 lbs per square uniformly over the entire surface.

Third: Finish the entire surface with a uniform coating of one of the surfacing options below.

Materials per 100 sq ft of roof area

FELTS:	J-M Centurian or Coated Asbestos Base Felt	1 layer
	J-M Asbestos Finishing Felt	2 layers
BITUMEN:	J-M 190 Asphalt (Slopes under 3" per foot)	46 lbs
	J-M 220 Asphalt (Slopes 3" or greater)	46 lbs
	Sprinkle mopping to plywood	10 lbs
SURFACING OPTIONS: (Select One)	J-M Topgard Type B — Fibrated	2-3 gals
	J-M Topgard Type F	1-1½ gals
	J-M Fibrated Aluminum Roof Coating	¾-1¼ gals
	J-M Topgard Type C	1 gal
	J-M Asphalt	20 lbs

Approximate applied Weight Min: 127 lbs. Max: 149 lbs.

Nailing

All nails or other fasteners are to be driven through tin caps unless the nail or fastener has an integral flat cap no less than 1" across.

BUILT-UP ROOFS by Johns-Manville/Greenwood Plaza, Denver, Colorado 80217

Regions 1, 2, & 3

*Specification **No. 103** for use over*

CONCRETE OR OTHER NON-NAILABLE DECKS

*on inclines of **1/2"** to **6"** per foot*

Johns-Manville
Smooth-Surface Asbestos BUILT-UP ROOFS

This specification is for use over any type of structural deck which is not nailable and which offers suitable surface to receive the roof. Poured and pre-cast concrete decks require priming. This specification is not to be used over lightweight insulating concrete decks either poured or pre-cast, or over fill made of lightweight insulating concrete.

Note: All information contained in "General Instructions" in the current Specification Manual for Johns-Manville Built-Up Roofs shall be considered part of this specification.

Flashings

See section on FLASHINGS, Specification Manual for J-M Built-Up Roofs.

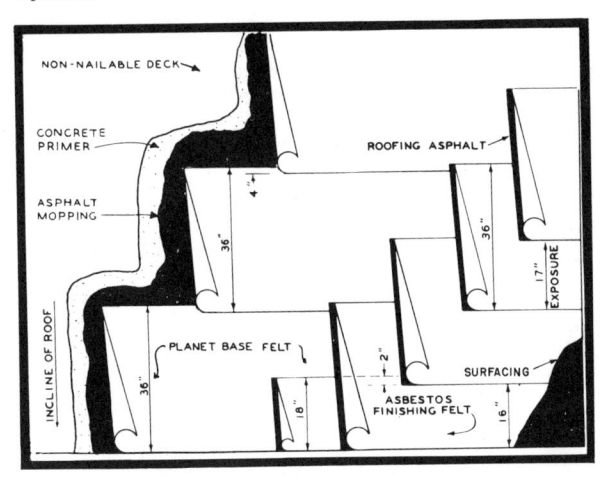

On slopes up to 2" apply finishing felts perpendicular to the slope starting at the low point of each slope. On slopes over 2" apply finishing felts parallel to the slope, nailing at the top of each run of felt on not over 9" centers. If run of felt exceeds 20' an additional line of nails shall be used at 20' intervals.

This specification is eligible for a 15-Year Guarantee only when, in the opinion of an authorized J-M Representative, all conditions listed in "General Instructions" of this Specification Manual have been met.

Application of Roofing

First: Regions 1 & 2: Use J-M Planet as the base felt, lapping each felt 4" over the preceding one and solidly mop the full width under each ply felt with asphalt using a minimum of 23 lbs. per square.

Region 3: Use J-M Planet as the base felt, lapping each felt 4" over the preceding one and spot mop the full width under each felt with asphalt using a minimum of 10 lbs. per square.

Second: Starting at the low edge (on slopes up to 2") apply one 18" wide, then over that one 36" wide J-M Asbestos Finishing Felt. Following felts are to be applied full width overlapping the preceding felt by 19" in such manner that at least 2 plies of felt cover the base felt at any point. Broom each felt so that it shall be firmly and uniformly set without voids into hot asphalt applied just before the felt at a minimum rate of 23 lbs. per square uniformly over the entire surface. On slopes over 2" per foot all felts shall be nailed at the top of each run of felt on not over 9" centers. If run of felt exceeds 20' an additional line of nails shall be used at 20' intervals.

Third: Finish the entire surface with a uniform coating of one of the surfacing options below.

Materials per 100 sq ft of roof area

CONCRETE PRIMER:	If required	1 gal
FELTS:	J-M Planet Base Felt	1 layer
	J-M Asbestos Finishing Felt	2 layers
BITUMEN:	J-M 190 Asphalt (Slopes under 3" per foot)	69 lbs.*
	J-M 220 Asphalt (Slopes 3" or greater)	69 lbs.*
SURFACING OPTIONS: (Select One)	J-M Topgard Type B — Fibrated	2-3 gals
	J-M Topgard Type F	1-1½ gals
	J-M Fibrated Aluminum Roof Coating	¾-1¼ gals
	J-M Topgard Type C	1 gal
	J-M Asphalt	20 lbs

Approximate applied Weight Min: 119 lbs. Max: 154 lbs.

*Deduct 13 lbs. if spot mopped.

Nailing

Where nailing is required, nailing strips must be provided. All nails or other fasteners are to be driven through tin caps unless the nail or fastener has an integral flat cap no less than 1" across.

BUILT-UP ROOFS by Johns-Manville/Greenwood Plaza, Denver, Colorado 80217

Regions 1, 2, & 3

Specification **No. 103-I** *for use over*

FESCO, FESCO-FOAM OR OTHER APPROVED INSULATION

*on inclines of **1/2"** to **6"** per foot*

Johns-Manville

Smooth-Surface Asbestos BUILT-UP ROOFS

This specification is for use over Fesco, Fesco-Foam or any type of approved insulation which is not nailable and which offers suitable surface to receive the roof. This specification is not to be used over light-weight aggregate concrete decks either poured or pre-cast, or over fill made of lightweight insulating concrete.

Note: All information contained in "General Instructions" in the current Specification Manual for Johns-Manville Built-Up Roofs shall be considered part of this specification.

Flashings

See section on FLASHINGS, Specification Manual for J-M Built-Up Roofs.

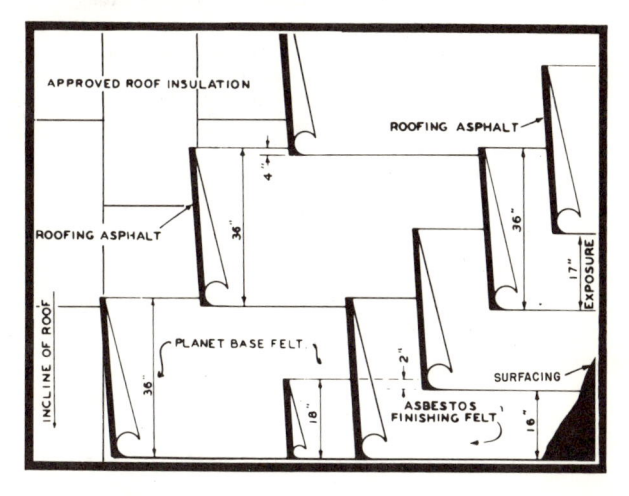

On slopes up to 2" apply finishing felts perpendicular to the slope starting at the low point of each slope. On slopes over 2" apply finishing felts parallel to the slope, nailing at the top of each run of felt on not over 9" centers. If run of felt exceeds 20' an additional line of nails shall be used at 20' intervals.

This specification is eligible for a 15-Year Guarantee only when, in the opinion of an authorized J-M Representative, all conditions listed in "General Instructions" of this Specification Manual have been met.

Application of Roofing

First: Apply one layer of J-M Planet Base Felt lapping each felt 4" over the preceding one. Mop the full width under each felt with asphalt using a minimum of 33 lbs. per square.

Second: Starting at the low edge (on slopes up to 2") apply one 18" wide, then over that one 36" wide J-M Asbestos Finishing Felt. Following felts are to be applied full width overlapping the preceding felt by 19" in such manner that at least 2 plies of felt cover the base felt at any point. Broom each felt so that it shall be firmly and uniformly set without voids into hot asphalt applied just before the felt at a minimum rate of 23 lbs. per square uniformly over the entire surface. On slopes over 2" per foot all felts shall be nailed at the top of each run of felt on not over 9" centers. If run of felt exceeds 20' an additional line of nails shall be used at 20' intervals.

Third: Finish the entire surface with a uniform coating of one of the surfacing options below.

Materials per 100 sq ft of roof area

FELTS:	J-M Planet Base Felt	1 layer
	J-M Asbestos Finishing Felt	2 layers
BITUMEN:	J-M 190 Asphalt (Slopes under 3" per foot)	79 lbs*
	J-M 220 Asphalt (Slopes 3" or greater)	79 lbs*
SURFACING OPTIONS: (Select One)	J-M Topgard Type B — Fibrated	2-3 gals
	J-M Topgard Type F	1-1½ gals
	J-M Fibrated Aluminum Roof Coating	¾-1¼ gals
	J-M Topgard Type C	1 gal
	J-M Asphalt	20 lbs

Approximate applied Weight Min: 152 lbs. Max: 187 lbs.

*If over Fesco Foam deduct 10 lbs.

Nailing

Where nailing is required, nailing strips must be provided. All nails or other fasteners are to be driven through tin caps unless the nail or fastener has an integral flat cap no less than 1" across.

BUILT-UP ROOFS by Johns-Manville/Greenwood Plaza, Denver, Colorado 80217

For Region 3 Only

Specification **No. 150** for use over

PLYWOOD OR OTHER NAILABLE DECKS

*on inclines of up to **1/2"** per foot*

Johns-Manville

Smooth-Surface Asbestos BUILT-UP ROOFS

This specification is to be used over any type of structural deck (without insulation) which can receive and adequately retain nails or other types of mechanical fasteners as may be recommended by the deck manufacturer. Examples of such decks are wood, plywood, and some lightweight aggregate concrete decks* no lighter than 32 PCF density, or that provide proper nail retention.

*Ventsulation Felt must be the base felt over these decks.

Note: All information contained in "General Instructions" in the current Specification Manual for Johns-Manville Built-Up Roofs shall be considered part of this specification.

Flashings

See section on FLASHINGS, Specification Manual for J-M Built-Up Roofs.

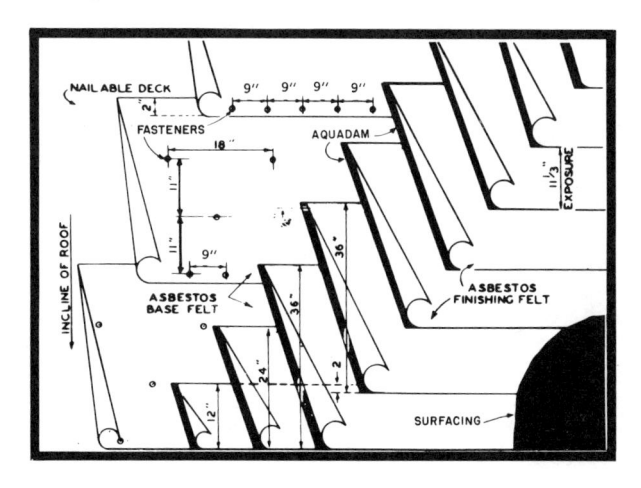

This specification is eligible for a 20-Year Guarantee only when, in the opinion of an authorized J-M Representative, all conditions listed in "General Instructions" of this Specification Manual have been met.

Application of Roofing

Over wood board decks one ply of sheathing paper must be used under the base felt next to the deck.

Regions 1 & 2: Use J-M Centurian as the base felt.

Region 3: Use either J-M Centurian or Coated Asbestos as the base felt.

First: Start at the low edge and working up the slope and perpendicular to the slope and lapping each felt 2" over the preceding one. Nail the laps at 9" centers and down the longitudinal center of each felt nail two rows of nails with rows spaced approximately 11" apart and nails staggered on approximately 18" centers. Use nails for fasteners appropriate to the type of deck.

Note: In Region 3 only if deck is plywood, base felt may be sprinkle mopped using 10 lbs. of asphalt per square.

Second: Starting at the low edge apply one 12" wide, then over that one 24" wide, then over both a full 36" wide J-M Asbestos Finishing Felt. Following felts are to be applied full width overlapping the preceding felt by 24⅔" in such manner that at least 3 plies of felt cover the base felt at any point. Broom each felt so that it shall be firmly and uniformly set without voids into hot J-M Aquadam applied just before the felt at a minimum rate of 23 lbs. per square uniformly over the entire surface.

Third: Finish the entire surface with a uniform coating of one of the surfacing options below.

Materials per 100 sq ft of roof area

FELTS:	J-M Centurian or Coated Asbestos Felt	1 layer
	J-M Asbestos Finishing Felt	3 layers
BITUMEN:	J-M Aquadam	69 lbs
	Sprinkle mopping to plywood	10 lbs
SURFACING OPTIONS: (Select One)	J-M Topgard Type F	1-1½ gals
	J-M Fibrated Aluminum Roof Coating	¾-1¼ gals
	J-M Aquadam	20 lbs

Approximate applied Weight Min: 147 lbs. Max: 182 lbs.

Nailing

All nails or other fasteners are to be driven through tin caps unless the nail or fastener has an integral flat cap no less than 1" across.

BUILT-UP ROOFS by Johns-Manville/Greenwood Plaza, Denver, Colorado 80217

Regions 1, 2, & 3

*Specification **No. 151** for use over*

CONCRETE OR OTHER NON-NAILABLE DECKS

*on inclines of up to **1/2"** per foot*

Johns-Manville
Smooth-Surface Asbestos BUILT-UP ROOFS

This specification is for use over any type of structural deck which is not nailable and which offers suitable surface to receive the roof. Poured and pre-cast concrete decks require priming. This specification is not to be used over lightweight insulating concrete decks either poured or pre-cast, or over fill made of lightweight insulating concrete.

Note: All information contained in "General Instructions" in the current Specification Manual for Johns-Manville Built-Up Roofs shall be considered part of this specification.

Flashings

See section on FLASHINGS, Specification Manual for J-M Built-Up Roofs.

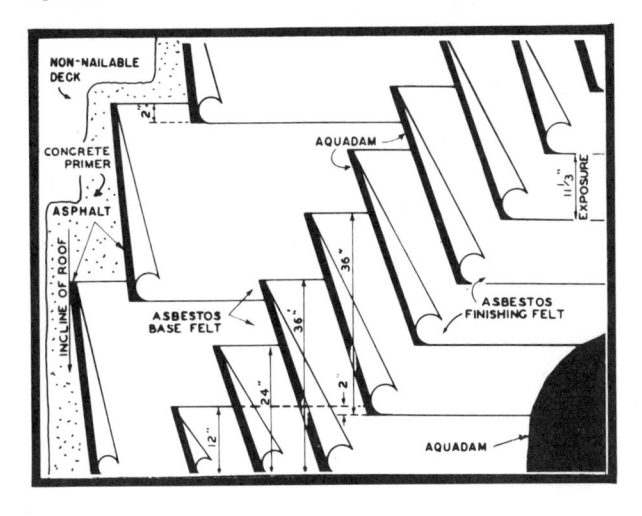

This specification is eligible for a 20-Year Guarantee only when, in the opinion of an authorized J-M Representative, all conditions listed in "General Instructions" of this Specification Manual have been met.

Application of Roofing

First: Regions 1 & 2: Use J-M Centurian as the base felt, lapping each felt 2" over the preceding one and solidly mop the full width under each ply felt with asphalt using a minimum of 23 lbs. per square.

Region 3: Use either J-M Centurian or Coated Asbestos as the base felt, lapping each felt 2" over the preceding one and spot mop the full width under each felt with asphalt using a minimum of 10 lbs. per square.

Second: Starting at the low edge apply one 12" wide, then over that one 24" wide, then over both a full 36" wide J-M Asbestos Finishing Felt. Following felts are to be applied full width overlapping the preceding felt by 24⅔" in such manner that at least 3 plies of felt cover the base felt at any point. Broom each felt so that it shall be firmly and uniformly set without voids into hot J-M Aquadam applied just before the felt at a minimum rate of 23 lbs. per square uniformly over the entire surface.

Third: Finish the entire surface with a uniform coating of one of the surfacing options below.

Materials per 100 sq. ft. of roof area

CONCRETE PRIMER: If required .... 1 gal.
FELTS: J-M Centurian or Coated Asbestos Base Felt ... 1 layer
        J-M Asbestos Finishing Felt .... 3 layers
BITUMEN:
        J-M 190 Asphalt .... 23 lbs.*
        J-M Aquadam .... 69 lbs.
SURFACING:
        J-M Topgard Type F .... 1-1½ gals.
        J-M Fibrated Aluminum Roof Coating ...... ¾-1¼ gals.
        J-M Aquadam .... 20 lbs.

Approximate Applied Weight Min: 168 lbs. Max: 205 lbs.

*Deduct 13 lbs. if spot mopped.

BUILT-UP ROOFS by Johns-Manville/Greenwood Plaza, Denver, Colorado 80217

r Regions 1, 2, & 3

Specification **No. 151-I** *for use over*

FESCO, FESCO-FOAM OR OTHER APPROVED INSULATION

*on inclines of up to **1/2"** per foot*

Johns-Manville
Smooth-Surface Asbestos BUILT-UP ROOFS

This specification is for use over Fesco, Fesco-Foam or any type of approved insulation which is not nailable and which offers suitable surface to receive the roof. This specification is not to be used over light-weight insulating concrete decks either poured or pre-cast, or over fill made of lightweight insulating concrete.

Note: All information contained in "General Instructions" in the current Specification Manual for Johns-Manville Built-Up Roofs shall be considered part of this specification.

Flashings

See section on FLASHINGS, Specification Manual for J-M Built-Up Roofs.

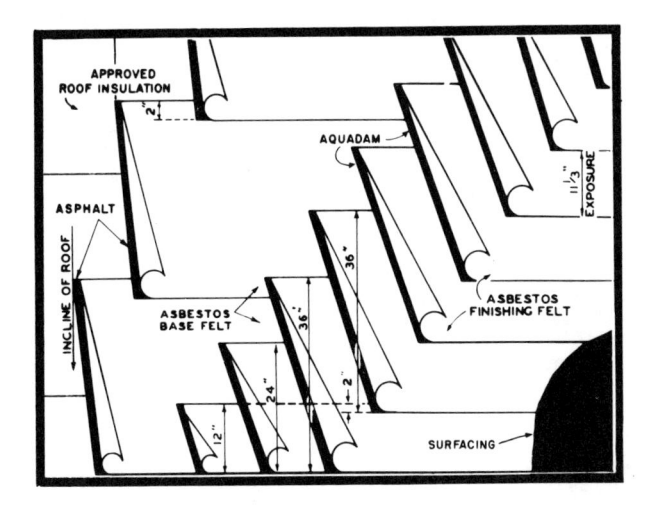

This specification is eligible for a 20-Year Guarantee only when, in the opinion of an authorized J-M Representative, all conditions listed in "General Instructions" of this Specification Manual have been met.

Application of Roofing

Regions 1 & 2: Use J-M Centurian as the base felt.

Region 3: Use either J-M Centurian or Coated Asbestos as the base felt.

First: Lap each felt 2" over the preceding one. Mop the full width under each felt with the appropriate hot J-M asphalt using a minimum of 33 lbs. per square.

Second: Starting at the low edge apply one 12" wide, then over that one 24" wide, then over both a full 36" wide J-M Asbestos Finishing Felt. Following felts are to be applied full width overlapping the preceding felt by 24⅔" in such manner that at least 3 plies of felt cover the base felt at any point. Broom each felt so that it shall be firmly and uniformly set without voids into hot J-M Aquadam applied just before the felt at a minimum rate of 23 lbs. per square uniformly over the entire surface.

Third: Finish the entire surface with a uniform coating of any of the surfacing options below.

Materials per 100 sq ft of roof area

FELTS:	J-M Centurian or Coated Asbestos Felt	1 layer
	J-M Asbestos Finishing Felt	3 layers
BITUMEN:	J-M 190 Asphalt	33 lbs.*
	J-M Aquadam	69 lbs.
SURFACING OPTIONS: (Select One)	J-M Topgard Type F	1-1½ gals
	J-M Fibrated Aluminum Roof Coating	¾-1¼ gals
	J-M Aquadam	20 lbs

*If over Fesco Foam deduct 10 lbs.

Approximate applied Weight Min: 170 lbs. Max: 182 lbs.

BUILT-UP ROOFS by Johns-Manville/Greenwood Plaza, Denver, Colorado 80217

Region 3 Only

Specification No. 154 for use over
CONCRETE OR OTHER NON-NAILABLE DECKS
on inclines of up to 1/2" per foot

Johns-Manville
Smooth-Surface Asbestos BUILT-UP ROOFS

This specification is for use over any type of structural deck which is not nailable and which offers suitable surface to receive the roof. Poured and pre-cast concrete decks require priming. This specification is not to be used over lightweight insulating concrete decks either poured or pre-cast, or over fill made of lightweight insulating concrete.

Note: All information contained in "General Instructions" in the current Specification Manual for Johns-Manville Built-Up Roofs shall be considered part of this specification.

Flashings

See section on FLASHINGS, Specification Manual for J-M Built-Up Roofs.

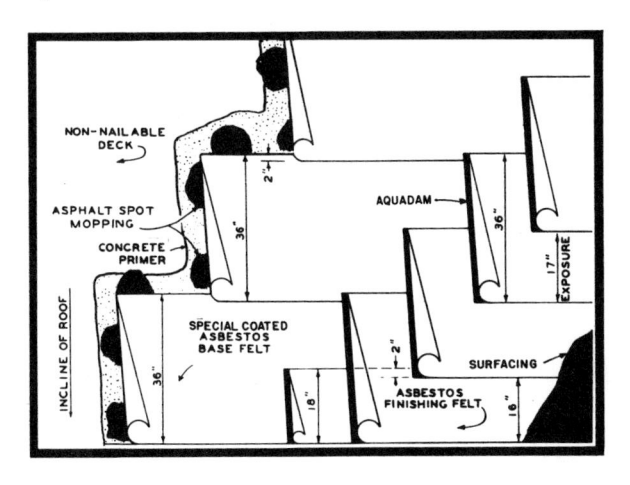

This specification is eligible for a 20-Year Guarantee only when, in the opinion of an authorized J-M Representative, all conditions listed in "General Instructions" of this Specification Manual have been met.

Application of Roofing

First: Apply one layer J-M Special Coated Asbestos Base Felt lapping each felt two inches over the preceding one. Spot mop the full width under each felt with 190-Asphalt using a minimum of 10 lbs per square.

Second: Starting at the low edge apply one 18" wide, then over that one 36" wide J-M Asbestos Finishing Felt. Following felts are to be applied full width overlapping the preceding felt by 19" in such manner that at least 2 plies of felt cover the base felt at any point. Broom each felt so that it shall be firmly and uniformly set without voids into hot Aquadam applied just before the felt at a minimum rate of 23 lbs. per square uniformly over the entire surface.

Third: Finish the entire surface with a uniform coating of one of the surfacing options below.

Materials per 100 sq ft of roof area

CONCRETE PRIMER:	If required	1 gal
FELTS:	J-M Special Coated Asbestos Felt	1 layer
	J-M Asbestos Finishing Felt	2 layers
BITUMEN:	J-M 190 Asphalt (spot mopping)	10 lbs
	J-M Aquadam	46 lbs
SURFACING OPTIONS: (Select One)	J-M Topgard Type F	1-1½ gals
	J-M Fibrated Aluminum Roof Coating	¾-1¼ gals
	J-M Aquadam	20 lbs

Approximate applied Weight Min: 149 lbs. Max: 164 lbs.

BUILT-UP ROOFS by Johns-Manville/Greenwood Plaza, Denver, Colorado 80217

or Region 3 Only

*Specification **No. 155-I** for use over*

FESCO, FESCO-FOAM OR OTHER APPROVED INSULATION

*on inclines of up to **1/2"** per foot*

Johns-Manville
Smooth-Surface Asbestos BUILT-UP ROOFS

This specification is for use over Fesco, Fesco-Foam or any type of approved insulation which is not nailable and which offers suitable surface to receive the roof. This specification is not to be used over light-weight aggregate concrete decks either poured or pre-cast, or over fill made of lightweight insulating concrete.

Note: All information contained in "General Instructions" in the current Specification Manual for Johns-Manville Built-Up Roofs shall be considered part of this specification.

Flashings

See section on FLASHINGS, Specification Manual for J-M Built-Up Roofs.

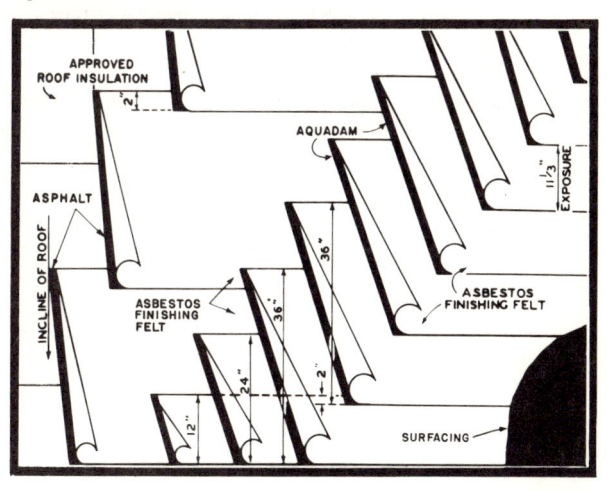

*This specification is eligible for a 20-Year Guarantee only when, in the opinion of an authorized J-M Representative, all conditions listed in "General Instructions" of this Specification **Manual** have been met.*

Application of Roofing

First: Apply one layer of J-M Asbestos Finishing Felt. Lap each felt 2" over the preceding one. Mop the full width under each felt with 190 Asphalt using a minimum of 33 lbs. per square.

Second: Starting at the low edge apply one 12" wide, then over that one 24" wide J-M Asbestos Finishing Felt, then over that a full 36" wide J-M Asbestos Finishing Felt. Following felts are to be applied full width overlapping the preceding felt by 24⅔" in such manner that at least 3 plies of felt cover the base felt at any point. Broom each felt so that it shall be firmly and uniformly set without voids into hot Aquadam applied just before the felt at a minimum rate of 23 lbs. per square uniformly over the entire surface.

Third: Finish the entire surface with a uniform coating of one of the surfacing options below.

Materials per 100 sq ft of roof area

FELTS:	J-M Asbestos Finishing Felt	4 layers
BITUMEN:	J-M 190 Asphalt	33 lbs.*
	J-M Aquadam	69 lbs
SURFACING OPTIONS: (Select One)	J-M Topgard Type F	1-1½ gals
	J-M Fibrated Aluminum Roof Coating	¾-1¼ gals
	J-M Aquadam	20 lbs

*If over Fesco Foam deduct 10 lbs.

Approximate applied Weight Min: 170 lbs. Max: 170 lbs.

BUILT-UP ROOFS by Johns-Manville/Greenwood Plaza, Denver, Colorado 80217

Regions 1, 2, & 3

*Specification **No. 152** for use over*

PLYWOOD OR OTHER NAILABLE DECKS

*on inclines of up to **1/2"** per foot*

Johns-Manville

Smooth-Surface Asbestos BUILT-UP ROOFS

This specification is to be used over any type of structural deck (without insulation) which can receive and adequately retain nails or other types of mechanical fasteners as may be recommended by the deck manufacturer. Examples of such decks are wood, plywood, and some lightweight aggregate concrete decks* no lighter than 32 PCF density, or that provide proper nail retention.

*Ventsulation Felt must be the base felt over these decks.

Note: All information contained in "General Instructions" in the current Specification Manual for Johns-Manville Built-Up Roofs shall be considered part of this specification.

Flashings
See section on FLASHINGS, Specification Manual for J-M Built-Up Roofs.

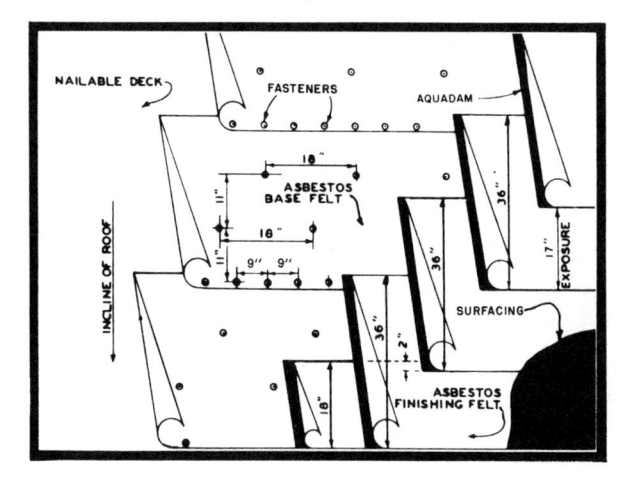

This specification is eligible for a 15-Year Guarantee when, in the opinion of an authorized J-M Representative, all conditions listed in "General Instructions" of this Specification Manual have been met.

Application of Roofing

Over wood board decks one ply of sheathing paper must be used under the base felt next to the deck.

Regions 1 & 2: Use J-M Centurian as the base felt.

Region 3: Use either J-M Centurian or Coated Asbestos as the base felt.

First: Start at the low edge and working up the slope and perpendicular to the slope and lapping each felt 2" over the preceding one. Nail the laps at 9" centers and down the longitudinal center of each felt nail two rows of nails with rows spaced approximately 11" apart and nails staggered on approximately 18" centers. Use nails for fasteners appropriate to the type of deck.

Note: In Region 3 only if deck is plywood, base felt may be sprinkle mopped using 10 lbs. of asphalt per square.

Second: Starting at the low edge apply one 18" wide, then over that one 36' wide J-M Asbestos Finishing Felt. Following felts are to be applied full width overlapping the preceding felt by 19" in such manner that at least 2 plies of felt cover the base felt at any point. Broom each felt so that it shall be firmly and uniformly set without voids into hot Aquadam applied just before the felt at a minimum rate of 23 lbs. per square uniformly over the entire surface.

Third: Finish the entire surface with a uniform coating of one of the surfacing options below.

Materials per 100 sq ft of roof area

FELTS:	J-M Centurian or Coated Asbestos Base Felt	1 layer
	J-M Asbestos Finishing Felt	2 layers
BITUMEN:	J-M Aquadam	46 lbs
	Sprinkle mopping to plywood	10 lbs
SURFACING OPTIONS: (Select One)	J-M Topgard Type F	1-1½ gals
	J-M Fibrated Aluminum Roof Coating	¾-1¼ gals
	J-M Aquadam	20 lbs

Approximate applied Weight Min: 127 lbs. Max: 149 lbs.

Nailing

All nails or other fasteners are to be driven through tin caps unless the nail or fastener has an integral flat cap no less than 1" across.

BUILT-UP ROOFS *by Johns-Manville/Greenwood Plaza, Denver, Colorado 80217*

*Specification **No. 153** for use over*

CONCRETE OR OTHER NON-NAILABLE DECKS

*on inclines of up to **1/2"** per foot*

Johns-Manville
Smooth-Surface Asbestos BUILT-UP ROOFS

This specification is for use over any type of structural deck which is not nailable and which offers suitable surface to receive the roof. Poured and pre-cast concrete decks require priming. This specification is not to be used over lightweight insulating concrete decks either poured or pre-cast, or over fill made of lightweight insulating concrete.

Note: All information contained in "General Instructions" in the current Specification Manual for Johns-Manville Built-Up Roofs shall be considered part of this specification.

Flashings

See section on FLASHINGS, Specification Manual for J M Built-Up Roofs.

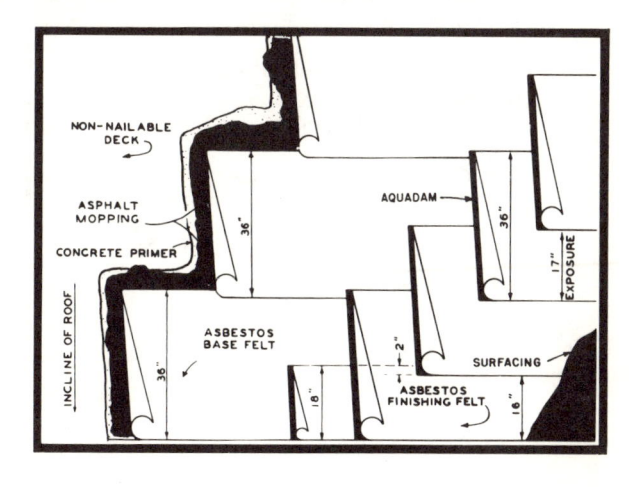

This specification is eligible for a 15-Year Guarantee only when, in the opinion of an authorized J-M Representative, all conditions listed in "General Instructions" of this Specification Manual have been met.

Application of Roofing

First: Regions 1 & 2: Use J-M Centurian as the base felt, lapping each felt 2" over the preceding one and solidly mop the full width under each ply felt with asphalt using a minimum of 23 lbs. per square.

Region 3: Use either J-M Centurian or Coated Asbestos as the base felt, lapping each felt 2" over the preceding one and spot mop the full width under each felt with asphalt using a minimum of 10 lbs. per square.

Second: Starting at the low edge apply one 18" wide, then over that one 36" wide J-M Asbestos Finishing Felt. Following felts are to be applied full width overlapping the preceding felt by 19" in such manner that at least 2 plies of felt cover the base felt at any point. Broom each felt so that it shall be firmly and uniformly set without voids into hot Aquadam applied just before the felt at a minimum rate of 23 lbs. per square uniformly over the entire surface.

Third: Finish the entire surface with a uniform coating of one of the surfacing options below.

Materials per 100 sq ft of roof area

CONCRETE PRIMER:	If required	1 gal
FELTS:	J-M Centurian or Coated Asbestos Base Felt	1 layer
	J-M Asbestos Finishing Felt	2 layers
BITUMEN:	J-M 190 Asphalt	23 lbs.*
	J-M Aquadam	46 lbs
SURFACING OPTIONS: (Select One)	J-M Topgard Type F	1-1½ gals
	J-M Fibrated Aluminum Roof Coating	¾-1¼ gals
	J-M Aquadam	20 lbs

Approximate applied Weight Min: 119 lbs. Max: 154 lbs.

*Deduct 13 lbs. if spot mopped.

BUILT-UP ROOFS by Johns-Manville/Greenwood Plaza, Denver, Colorado 80217

Specification **No. 153-I** *for use over*

FESCO, FESCO-FOAM OR OTHER APPROVED INSULATION

on inclines of up to 1/2" per foot

Johns-Manville
Smooth-Surface Asbestos BUILT-UP ROOFS

This specification is for use over Fesco, Fesco-Foam or any type of approved insulation which is not nailable and which offers suitable surface to receive the roof. This specification is not to be used over light-weight aggregate concrete decks either poured or pre-cast, or over fill made of lightweight insulation concrete.

Note: All information contained in "General Instructions" in the current Specification Manual for Johns-Manville Built-Up Roofs shall be considered part of this specification.

Flashings

See section on FLASHINGS, Specification Manual for J-M Built-Up Roofs.

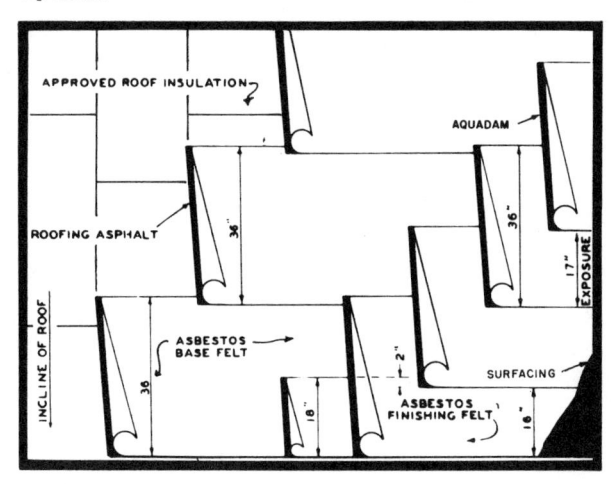

This specification is eligible for a 15-Year Guarantee only when, in the opinion of an authorized J-M Representative, all conditions listed in "General Instructions" of this Specification Manual have been met.

Application of Roofing

Regions 1 & 2: Use J-M Centurian as the base felt.

Region 3: Use either J-M Centurian or Coated Asbestos as the base felt.

First: Lap each felt 2" over the preceding one. Mop the full width under each felt with the appropriate hot J-M asphalt using a minimum of 33 lbs. per square.

Second: Starting at the low edge apply one 18" wide, then over that one 36" wide J-M Asbestos Finishing Felt. Following felts are to be applied full width overlapping the preceding felt by 19" in such manner that at least 2 plies of felt cover the base felt at any point. Broom each felt so that it shall be firmly and uniformly set without voids into hot Aquadam applied just before the felt at a minimum rate of 23 lbs. per square uniformly over the entire surface.

Third: Finish the entire surface with a uniform coating of any of the surfacing options below.

Materials per 100 sq ft of roof area

FELTS:	J-M Centurian or Coated Asbestos Felt	1 layer
	J-M Asbestos Finishing Felt	2 layers
BITUMEN:	J-M 190 Asphalt	33 lbs*
	J-M Aquadam	46 lbs.
SURFACING OPTIONS: (Select One)	J-M Topgard Type F	1-1½ gals
	J-M Fibrated Aluminum Roof Coating	¾-1¼ gals
	J-M Aquadam	20 lbs

*If over Fesco Foam deduct 10 lbs.

Approximate applied Weight Min: 127 lbs. Max: 144 lbs.

BUILT-UP ROOFS by Johns-Manville/Greenwood Plaza, Denver, Colorado 80217

Region 3 Only

*Specification **No. 156-I** for use over*

FESCO, FESCO-FOAM OR OTHER APPROVED INSULATION

*on inclines of up to **1/2"** per foot*

Johns-Manville
Smooth-Surface Asbestos BUILT-UP ROOFS

This specification is for use over Fesco, Fesco-Foam or any type of approved insulation which is not nailable and which offers suitable surface to receive the roof. This specification is not to be used over light-weight aggregate concrete decks either poured or pre-cast, or over fill made of lightweight insulation concrete.

Note: All information contained in "General Instructions" in the current Specification Manual for Johns-Manville Built-Up Roofs shall be considered part of this specification.

Flashings

See section on FLASHINGS, Specification Manual for J-M Built-Up Roofs.

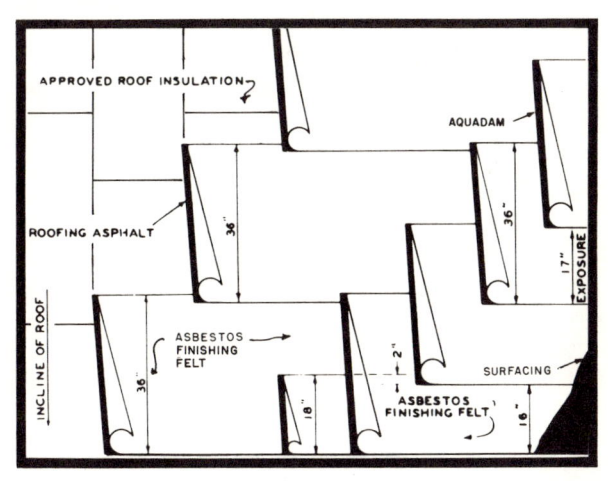

This specification is eligible for a 15-Year Guarantee only when, in the opinion of an authorized J-M Representative, all conditions listed in "General Instructions" of this Specification Manual have been met.

Application of Roofing

First: Apply one layer of J-M Asbestos Finishing Felt lapping each felt 2″ over the preceding one. Mop the full width under each felt with 190 Asphalt using a minimum of 33 lbs. per square.

Second: Starting at the low edge apply one 18″ wide, then over that one 36″ wide J-M Asbestos Finishing Felt. Following felts are to be applied full width overlapping the preceding felt by 19″ in such manner that at least 2 plies of felt cover the base felt at any point. Broom each felt so that it shall be firmly and uniformly set without voids into hot Aquadam applied just before the felt at a minimum rate of 23 lbs. per square uniformly over the entire surface.

Third: Finish the entire surface with a uniform coating of one of the surfacing options below.

Materials per 100 sq. ft. of roof area

FELTS: J-M Asbestos Finishing Felt .... 3 layers
BITUMEN:
        J-M 190 Asphalt .... 33 lbs *
        J-M Aquadam .... 46 lbs
SURFACING:
        J-M Topgard Type F .... 1-1½ gals.
        J-M Fibrated Aluminum Roof Coating ...... ¾-1¼ gals.
        J-M Asphalt .... 20 lbs.

*If over Fesco Foam deduct 10 lbs.

Approximate applied Weight Min: 132 lbs. Max: 132 lbs.

BUILT-UP ROOFS *by Johns-Manville/Greenwood Plaza, Denver, Colorado 80217*

For Regions 1, 2, & 3

*Specification **No. 2000** for use over*
PLYWOOD OR OTHER NAILABLE DECKS
*on inclines of **1/2"** to **6"** per foot*

Johns-Manville
Smooth-Surface Asbestos
COLD APPLICATION BUILT-UP ROOFS

This specification is to be used over any type of structural deck (without insulation) which can receive and adequately retain nails or other types of mechanical fasteners as may be recommended by the deck manufacturer. Examples of such decks are wood, plywood, and some lightweight aggregate concrete decks* no lighter than 32 PCF density, or that provide proper nail retention.

*Ventsulation Felt must be the base felt over these decks.

Note: All information contained in "General Instructions" in the current Specification Manual for Johns-Manville Built-Up Roofs shall be considered part of this specification.

Flashings
See section on FLASHINGS, Specification Manual for J-M Built-Up Roofs.

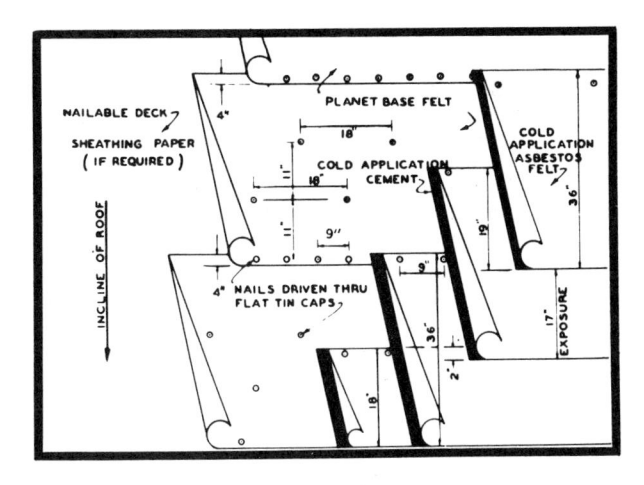

75

Application of Roofing

Over wood board decks one ply of sheathing paper must be used under the base sheet next to the deck.

Over plywood, sprinkle cement base felt to the deck. Omit nailing if slope allows, in Region 3 only.

First: Apply one layer of J-M Planet Base Felt starting at the low edge and working up the slope and perpendicular to the slope, and lapping each course 4" over the preceding one. Nail the laps at 9" centers and down the longitudinal center of each felt nail two rows of nails with the rows spaced approximately 11" apart and nails staggered on approximately 18" centers. Use nails or fasteners appropriate to the type of deck.

Second: Starting at the low edge apply one 18" wide, then over that one full 36" wide, J-M Cold Application Felt. Following felts are to be applied full width overlapping the preceding felt by 19" in such manner that at least 2 plies of felt cover the base felt at any point. On slopes 2" per foot or greater nail each felt at approximately 9" centers adjacent to the back edge. Broom each felt so that it shall be firmly and uniformly set without voids into J-M Cold Application Cement applied just before the felt at a minimum rate of approximately 2 gallons per square uniformly over the entire surface.

Materials per 100 sq ft of roof area

SHEATHING PAPER: Wood board decks only	1 layer
FELTS: J-M Planet Base Felt	1 layer
J-M Cold Application Asbestos Felt	2 layers
CEMENT: Cold Application Cement	4 gal

Approximate Installed Weight Min: 145 lbs. Max: 150 lbs.

Nailing

All nails or other fasteners are to be driven through tin caps unless the nail or fastener has an integral flat cap no less than 1" across.

Cement

Cold Application cement shall be used as received in the original container. It shall be spread evenly with a three or four knot roofing brush. Felts shall be rolled immediately into the cement and embedded solidly leaving no voids or air pockets. The surface of the felt shall be worked with a broom or squeegee to assure complete contact.

As the cement does not set up until the solvent has volatilized, mechanics and others required to traverse the roof during and within two to three days after application shall exercise caution to prevent damage.

BUILT-UP ROOFS *by Johns-Manville/Greenwood Plaza, Denver, Colorado 80217*

For Regions 1, 2, & 3

Specification No. 2001 for use over

NON-NAILABLE DECKS or APPROVED INSULATION

on inclines of 1/2" to 6" per foot

Johns-Manville

Smooth-Surface Asbestos
COLD APPLICATION BUILT-UP ROOFS

This specification is for use over any type of structural deck or approved insulation which is not nailable and which offers suitable surface to receive the roof. Poured and pre-cast concrete decks require priming. This specification is not to be used over lightweight aggregate concrete decks either poured or pre-cast, or over fill made of lightweight aggregate concrete.

Note: All information contained in "General Instructions" in the current Specification Manual for Johns-Manville Built-Up Roofs shall be considered part of this specification.

Flashings
See section on FLASHINGS, Specification Manual for J-M Built-Up Roofs.

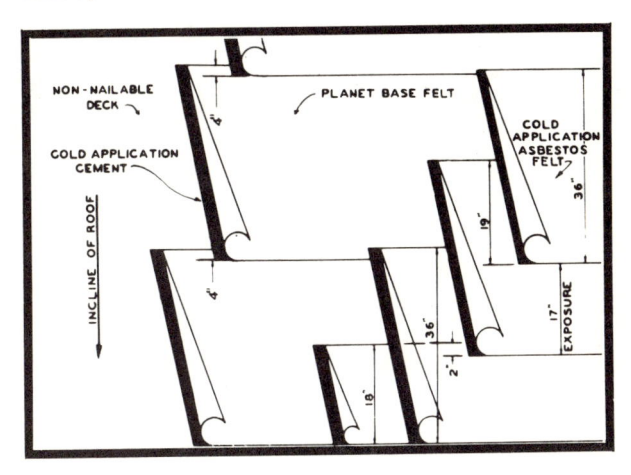

Application of Roofing

First: Apply one layer of J-M Planet Base Felt lapping each course four inches over the preceding one. Cement the full width under each felt with J-M Cold Application Cement using approximately 3 gallons per square.

Second: Starting at the low edge apply one 18" wide, then over that one 36" wide J-M Cold Application Asbestos Felt. Following felts are to be applied full width overlapping the preceding felt by 19" in such manner that at least 2 plies of felt cover the base felt at any point. Broom each felt so that it shall be firmly and uniformly set without voids into J-M Cold Application Cement applied just before the felt at a rate of approximately 2 gallons per square uniformly over the entire surface.

Materials per 100 sq ft of roof area

CONCRETE PRIMER: If required	1 gal
FELTS: J-M Planet Base Felt	1 layer
J-M Cold Application Asbestos Felt	2 layers
CEMENT: Cold Application Cement	7 gal

Approximate Installed Weight Min: 169 lbs. Max: 169 lbs.

Cement

Cold Application cement shall be used as received in the original container. It shall be spread evenly with a three or four knot roofing brush. Felts shall be rolled immediately into the cement and embedded solidly leaving no voids or air pockets. The surface of the felt shall be worked with a broom or squeegee to assure complete contact.

As the cement does not set up until the solvent has volatilized, mechanics and others required to traverse the roof during and within two to three days after application shall exercise caution to prevent damage.

BUILT-UP ROOFS *by Johns-Manville / P.O. Box 5108, Denver, Colorado 80217*

Johns-Manville
GlasKap & Flexstone
MINERAL SURFACE BUILT-UP ROOFS

MINERAL SURFACE

Johns-Manville
GlasKap & Flexstone
MINERAL SURFACE CAP SHEETS

Mineral surfaced cap sheets form the last, visible ply in the roof membrane. The totally inorganic mat is saturated and coated with weathering-grade asphalt into which is embedded opaque, non-combustible mineral granules on the exposed surface. The resulting sheet yields a roof in white or a choice of colors that enhances the appearance of the building it protects.

Regions 1, 2, & 3

Specification **No. 400** *for use over*

PLYWOOD OR OTHER NAILABLE DECKS

on inclines of **1/4"** *to* **6"** *per foot*

Johns-Manville
Mineral-Surface BUILT-UP ROOFS

This specification is to be used over any type of structural deck (without insulation) which can receive and adequately retain nails or other types of mechanical fasteners as may be recommended by the deck manufacturer. Examples of such decks are wood, plywood, and some lightweight aggregate concrete decks* no lighter than 32 PCF density, or that provide proper nail retention.

*Ventsulation Felt must be the base felt over these decks.

Note: All information contained in "General Instructions" in the current Specification Manual for Johns-Manville Built-Up Roofs shall be considered part of this specification.

Flashings
See section on FLASHINGS, Specification Manual for J-M Built-Up Roofs.

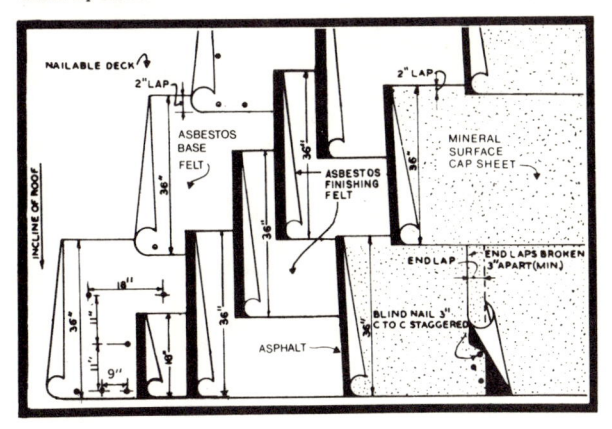

This specification is eligible for a 20-Year Guarantee only when, in the opinion of an authorized J-M Representative, all conditions listed in "General Instructions" of this Specification Manual have been met.

Application of Roofing

Over wood board decks one ply of sheathing paper must be used under the base felt next to the deck.

Regions 1 & 2: Use J-M Centurian as the base felt.

Region 3: Use either J-M Centurian or Coated Asbestos as the base felt.

First: Start at the low edge and working up the slope and perpendicular to the slope and lapping each felt 2" over the preceding one. Nail the laps at 9" centers and down the longitudinal center of each felt nail two rows of nails with rows spaced approximately 11" apart and nails staggered on approximately 18" centers. Use nails for fasteners appropriate to the type of deck.

Note: In Region 3 only if deck is plywood, base felt may be sprinkle mopped using 10 lbs. of asphalt per square.

Second: Apply two layers of J-M Asbestos Finishing Felt. lapping each felt 19" over the preceding one. Mop the full width of each felt with asphalt at a minimum rate of 23 lbs. per square. Broom each felt so that it shall be firmly and uniformly set without voids into the hot asphalt. On slopes 2" per ft. or greater nail each felt at approximately 9" centers adjacent to the back edge.

Third: Starting at the low edge apply one layer of J-M Cap Sheet lapping each layer 2" over the preceding one. Lap the end 6" over the preceding felt. Mop the full width under each layer with the asphalt at a minimum rate of 23 lbs. per square making sure that all edges are well sealed, and with the Cap Sheet uniformly set without voids into the hot asphalt. On slopes 2" per foot and greater nail adjacent to the top edge on 12" centers. Nail laps on 3" centers and mop the full width with asphalt.

Fourth: In areas of standing water over the Cap Sheet with slopes of ¼" or less, apply hot asphalt at a rate of about 50 lbs. per square and embed therein approximately 300 lbs. of gravel or slag.

Materials per 100 sq. ft. of roof area

FELTS: J-M Centurian or Coated Asbestos Base Felt .... 1 layer
J-M Asbestos Finishing Felt .... 2 layers

REGIONS 1 & 2
J-M GlasKap Mineral Surface Fiber Glass
Cap Sheet .... 1 layer

REGION 3
J-M Flexstone Mineral Surface Cap Sheet or
J-M GlasKap Mineral Surface Fiber
Glass Cap Sheet .... 1 layer

BITUMEN:
J-M 190 Asphalt (Slopes under 3" per foot) ..... 69 lbs.
J-M 220 Asphalt (Slopes 3" or greater) .... 69 lbs.
Sprinkle mopping to plywood .... 10 lbs.

Approximate applied Weight Min: 183 lbs. Max: 219 lbs.

Nailing

All nails or other fasteners are to be driven through tin caps unless the nail or fastener has an integral flat cap no less than 1" across.

Note: Before starting application of the J-M Flexstone Cap Sheet cut into lengths of 12' to 18' and allow to flatten.

BUILT-UP ROOFS by Johns-Manville / P.O. Box 5108, Denver, Colorado 80217

or Regions 1, 2, & 3

Specification **No. 401** *for use over*

CONCRETE OR OTHER NON-NAILABLE DECKS

on inclines of **1/4"** *to* **6"** *per foot*

Johns-Manville
Mineral-Surface BUILT-UP ROOFS

This specification is for use over any type of structural deck which is not nailable and which offers suitable surface to receive the roof. Poured and pre-cast concrete decks require priming. This specification is not to be used over lightweight insulating concrete decks either poured or pre-cast, or over fill made of lightweight insulating concrete.

Note: All information contained in "General Instructions" in the current Specification Manual for Johns-Manville Built-Up Roofs shall be considered part of this specification.

Flashings

See section on FLASHINGS, Specification Manual for J-M Built-Up Roofs.

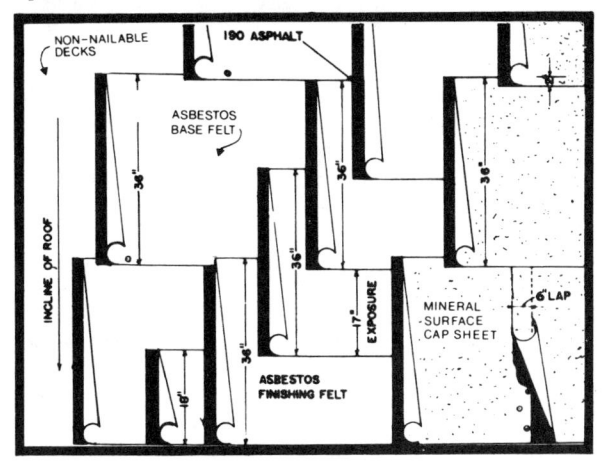

On slopes up to 2" apply finishing felts perpendicular to the slope starting at the low point of each slope. On slopes over 2" apply finishing felts parallel to the slope, nailing at the top of each run of felt on not over 9" centers. If run of felt exceeds 20' an additional line of nails shall be used at 20' intervals.

This specification is eligible for a 20-Year Guarantee only when, in the opinion of an authorized J-M Representative, all conditions listed in "General Instructions" of this Specification Manual have been met.

Application of Roofing

First: Regions 1 & 2: Use J-M Centurian as the base felt, lapping each felt 2" over the preceding one and solidly mop the full width under each ply felt with asphalt using a minimum of 23 lbs. per square.

Region 3: Use either J-M Centurian or Coated Asbestos as the base felt, lapping each felt 2" over the preceding one and spot mop the full width under each felt with asphalt using a minimum of 10 lbs. per square.

Second: Apply two layers of J-M Asbestos Finishing Felt, lapping each felt 19" over the preceding one. Mop the full width under each felt with asphalt at a minimum rate of 23 lbs. per square per ply. Broom each felt so that it shall be firmly and uniformly set without voids into the hot asphalt.

Third: Starting at the low edge (on slopes up to 2") or at the side opposite the prevailing wind (on slopes over 2"), apply one layer of Cap Sheet, lapping each layer 2" over the preceding one. Lap the end 6" over the preceding felt. Mop the full width under each layer with the asphalt at a minimum rate of 23 lbs. per square making sure that all edges are well sealed, and with the J-M Cap Sheet uniformly set without voids into the hot asphalt.

Fourth: In areas of standing water over the Cap Sheet with slopes of ¼" or less, apply hot asphalt at a rate of about 50 lbs. per square and embed therein approximately 300 lbs. of gravel or slag.

Materials per 100 sq. ft. of roof area

CONCRETE
PRIMER:
        If required .... 1 gal.

FELTS: J-M Centurian or Coated Asbestos Base Felt .... 1 layer
        J-M Asbestos Finishing Felt .... 2 layers

        REGIONS 1 & 2
        J-M GlasKap Mineral Surface Fiber Glass
          Cap Sheet .... 1 layer

        REGION 3
        J-M Flexstone Mineral Surface Cap Sheet or
        J-M GlasKap Mineral Surface Fiber
          Glass Cap Sheet .... 1 layer

BITUMEN:
        J-M 190 Asphalt (Slopes under 3" per foot) .... 92 lbs.*
        J-M 220 Asphalt (Slopes 3" or greater) .... 92 lbs.*

Approximate applied Weight Min: 219 lbs. Max: 242 lbs.

*Deduct 13 lbs. if Spot Mopped.

Nailing

Where nailing is required, nailing strips must be provided. All nails or other fasteners are to be driven through tin caps unless the nail or fastener has an integral flat cap no less than 1" across.

Note: Before starting application of the J-M Flexstone Cap Sheet cut into lengths of 12' to 18' and allow to flatten.

BUILT-UP ROOFS by Johns-Manville / P.O. Box 5108, Denver, Colorado 80217

or Regions 1, 2, & 3

*Specification **No. 401-I** for use over*

FESCO, FESCO-FOAM OR OTHER APPROVED INSULATION

*on inclines of **1/4"** to **6"** per foot*

Johns-Manville
Mineral-Surface BUILT-UP ROOFS

This specification is for use over Fesco, Fesco-Foam or any type of approved insulation which is not nailable and which offers suitable surface to receive the roof. This specification is not to be used over light-weight insulating concrete decks either poured or pre-cast, or over fill made of lightweight insulating concrete.

Note: All information contained in "General Instructions" in the current Specification Manual for Johns-Manville Built-Up Roofs shall be considered part of this specification.

Flashings

See section on FLASHINGS, Specification Manual for J-M Built-Up Roofs.

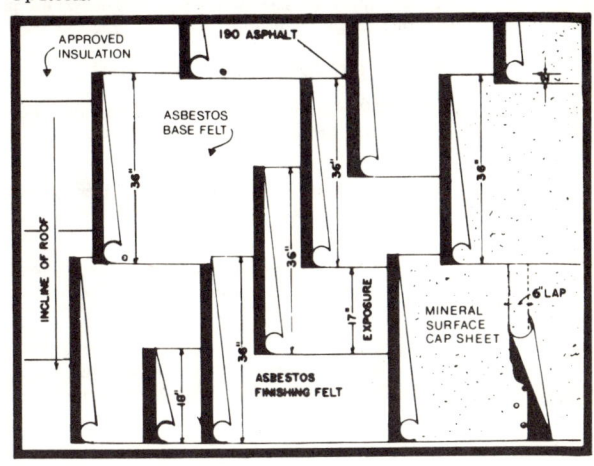

On slopes up to 2" apply felts perpendicular to the slope starting at the low point of each slope. On slopes over 2" apply felts parallel to the slope, nailing at the top of each run of felt on not over 9" centers. If run of felt exceeds 20' an additional line of nails shall be used at 20' intervals.

This specification is eligible for a 20-Year Guarantee only when, in the opinion of an authorized J-M Representative, all conditions listed in "General Instructions" of this Specification Manual have been met.

Application of Roofing

Regions 1 & 2: Use J-M Centurian as the base felt.

Region 3: Use either J-M Centurian or Coated Asbestos as the base felt.

First: Lap each felt 2" over the preceding one. Mop the full width under each felt with the appropriate hot J-M asphalt using a minimum of 33 lbs. per square.

Second: Apply two layers of J-M Asbestos Finishing Felt, lapping each felt 19" over the preceding one. Mop the full width under each felt with asphalt at a minimum rate of 23 lbs. per square per ply. Broom each felt so that it shall be firmly and uniformly set without voids into the hot asphalt.

Third: Starting at the low edge (on slopes up to 2") or at the side opposite the prevailing wind (on slopes over 2"), apply one layer of Cap Sheet, lapping each layer 2" over the preceding one. Lap the end 6" over the preceding felt. Mop the full width under each layer with the asphalt at a minimum rate of 23 lbs. per square making sure that all edges are well sealed, and with the J-M Cap Sheet uniformly set without voids into the hot asphalt.

Fourth: In areas of standing water over the Cap Sheet with slopes of ¼" or less, apply hot asphalt at a rate of about 50 lbs. per square and embed therein approximately 300 lbs. of gravel or slag.

Materials per 100 sq. ft. of roof area

FELTS: J-M Centurian or Coated Asbestos Base Felt .... 1 layer
J-M Asbestos Finishing Felt .... 2 layers

REGIONS 1 & 2
J-M GlasKap Mineral Surface Fiber Glass
  Cap Sheet .... 1 layer

REGION 3
J-M Flexstone Mineral Surface Cap Sheet or
J-M GlasKap Mineral Surface Fiber
  Glass Cap Sheet .... 1 layer

BITUMEN:
J-M 190 Asphalt (Slopes under 3" per foot) ... 102 lbs.*
J-M 220 Asphalt (Slopes 3" or greater) .... 102 lbs.*

Approximate applied Weight Min: 229 lbs. Max: 252 lbs.

*Deduct 10 lbs. over Fesco-Foam.

Nailing

Where nailing is required, nailing strips must be provided. All nails or other fasteners are to be driven through tin caps unless the nail or fastener has an integral flat cap no less than 1" across.

Note: Before starting application of the J-M Flexstone Cap Sheet cut into lengths of 12' to 18' and allow to flatten.

BUILT-UP ROOFS by Johns-Manville / P.O. Box 5108, Denver, Colorado 80217

Regions 1, 2, & 3

*Specification **No. 402** for use over*

PLYWOOD OR OTHER NAILABLE DECKS

*on inclines of **1/4"** to **6"** per foot*

Johns-Manville
Mineral-Surface BUILT-UP ROOFS

This specification is to be used over any type of structural deck (without insulation) which can receive and adequately retain nails or other types of mechanical fasteners as may be recommended by the deck manufacturer. Examples of such decks are wood, plywood, and some lightweight aggregate concrete decks* no lighter than 32 PCF density, or that provide proper nail retention.

*Ventsulation Felt must be the base felt over these decks.

Note: All information contained in "General Instructions" in the current Specification Manual for Johns-Manville Built-Up Roofs shall be considered part of this specification.

Flashings
See section on FLASHINGS, Specification Manual for J-M Built-Up Roofs.

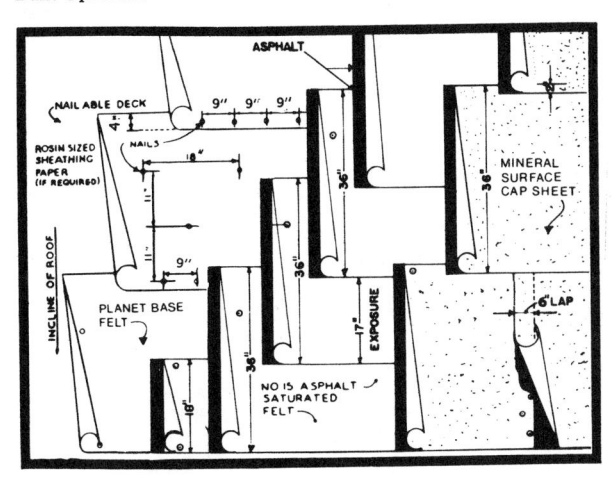

This specification is eligible for a 15-Year Guarantee only when, in the opinion of an authorized J-M Representative, all conditions listed in "General Instructions" of this Specification Manual have been met.

Application of Roofing

Over wood board decks one ply of sheathing paper must be used under the base felt next to the deck.

Over plywood, omit sheathing paper and sprinkle mop base felt to the deck, omit nailing if slope allows, in Region 3 only.

First: Apply one layer of J-M Planet Base Felt starting at the low edge either parallel or perpendicular to the slope, and lapping each course 4" over the preceding one. Nail the laps at 9" centers and down the longitudinal center of each felt nail two rows of nails with the rows spaced approximately 11" apart and nails staggered on approximately 18" centers. Use nails or fasteners appropriate to the type of deck.

Second: Apply two layers of J-M Asphalt Saturated Felt (perforated), lapping each felt 19" over the preceding one. Mop the full width under each felt with asphalt at a minimum rate of 23 lbs. per square. Broom each felt so that it shall be firmly and uniformly set without voids into the hot asphalt. On slopes 2" per foot or greater nail each felt at approximately 9" centers adjacent to the back edge.

Third: Starting at the low edge (on slopes up to 2") or at the side opposite the prevailing wind (on slopes over 2"), apply one layer of J-M Cap Sheet, lapping each layer 2" over the preceding one. Lap the end 6" over the preceding felt. Mop the full width under each layer with the asphalt at a minimum rate of 23 lbs. per square making sure that all edges are well sealed, and with the J-M Cap Sheet uniformly set without voids into the hot asphalt. Back-nail if required.

Fourth: In areas of standing water over the Cap Sheet with slopes of $1/4"$ or less, apply hot asphalt at a rate of about 50 lbs. per square and embed therein approximately 300 lbs. of gravel or slag.

Materials per 100 sq. ft. of roof area

FELTS:	J-M Planet Base Sheet	1 layer
	J-M Asphalt Saturated Felt	2 layers
	REGIONS 1 & 2 J-M GlasKap Mineral Surface Fiber Glass Cap Sheet	1 layer
	REGION 3 J-M Flexstone Mineral Surface Cap Sheet or J-M GlasKap Mineral Surface Fiber Glass Cap Sheet	1 layer
BITUMEN:		
	J-M 190 Asphalt (Slopes under 3" per foot)	69 lbs.
	J-M 220 Asphalt (Slopes 3" or greater)	69 lbs.
	Sprinkle mopping to plywood	10 lbs.

Approximate applied Weight Min: 201 lbs. Max: 219 lbs.

Nailing

All nails or other fasteners are to be driven through tin caps unless the nail or fastener has an integral flat cap no less than 1" across.

Note: Before starting application of the J-M Flexstone Cap Sheet cut into lengths of 12' to 18' and allow to flatten.

BUILT-UP ROOFS by Johns-Manville / P.O. Box 5108, Denver, Colorado 80217

Regions 1, 2, & 3

*Specification **No. 403** for use over*

CONCRETE OR OTHER NON-NAILABLE DECKS

*on inclines of **1/4"** to **6"** per foot*

Johns-Manville

Mineral-Surface BUILT-UP ROOFS

This specification is for use over any type of structural deck which is not nailable and which offers suitable surface to receive the roof. Poured and pre-cast concrete decks require priming. This specification is not to be used over lightweight insulating concrete decks either poured or pre-cast, or over fill made of lightweight insulating concrete.

Note: All information contained in "General Instructions" in the current Specification Manual for Johns-Manville Built-Up Roofs shall be considered part of this specification.

Flashings

See the section titled FLASHINGS, in the Specification Manual for J-M Built-Up Roofs.

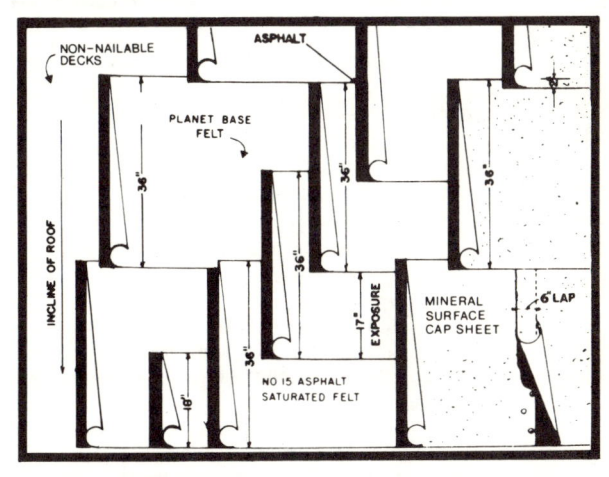

On slopes up to 2" apply felts perpendicular to the slope starting at the low point of each slope. On slopes over 2" apply felts parallel to the slope, nailing at the top of each run of felt on not over 9" centers. If run of felt exceeds 20' an additional line of nails shall be used at 20' intervals.

This specification is eligible for a 15-Year Guarantee only when, in the opinion of an authorized J-M Representative, all conditions listed in "General Instructions" of this Specification Manual have been met.

Application of Roofing

First: Regions 1 & 2: Use J-M Planet as the base felt, lapping each felt 4" over the preceding one and solidly mop the full width under each ply felt with asphalt using a minimum of 23 lbs. per square.

Region 3: Use J-M Planet as the base felt, lapping each felt 4" over the preceding one and spot mop the full width under each felt with asphalt using a minimum of 10 lbs. per square.

Second: Apply two layers of J-M No. 15 Asphalt Saturated Felt (perforated), lapping each felt 19" over the preceding one. Mop the full width of each felt with asphalt at a minimum rate of 23 lbs. per square. Broom each felt so that it shall be firmly and uniformly set without voids into the hot asphalt. On slopes 2" per ft. or greater nail each felt at approximately 9" centers adjacent to the back edge.

Third: Starting at the low edge (on slopes up to 2") or at the side opposite the prevailing wind (on slopes over 2"), apply one layer of J-M Cap Sheet, lapping each layer 2" over the edge of the preceding one. Lap the end 6" over the preceding felt. Mop the full width under each layer with the asphalt, making sure that all edges are well sealed, and with the J-M Cap Sheet uniformly set without voids into the hot asphalt. Back-nail if required.

Fourth: In areas of standing water over the Cap Sheet with slopes of 1/4" or less, apply hot asphalt at a rate of about 50 lbs. per square and embed therein approximately 300 lbs. of gravel or slag.

Materials per 100 sq. ft. of roof area

CONCRETE
PRIMER:
       If required .... 1 gal.

FELTS: J-M Planet Base Sheet .... 1 layer
          J-M Asphalt Saturated Felt .... 2 layers

       REGIONS 1 & 2
       J-M GlasKap Mineral Surface Fiber Glass
       Cap Sheet .... 1 layer

       REGION 3
       J-M Flexstone Mineral Surface Cap Sheet or
       J-M GlasKap Mineral Surface Fiber
       Glass Cap Sheet .... 1 layer

BITUMEN:
        J-M 190 Asphalt (Slopes under 3" per foot) .... 92 lbs.*
        J-M 220 Asphalt (Slopes 3" or greater) .... 92 lbs.*

Approximate applied Weight Min: 237 lbs. Max: 249 lbs.

*Deduct 13 lbs. if Spot Mopped.

Nailing

Where nailing is required, nailing strips must be provided. All nails or other fasteners are to be driven through tin caps unless the nail or fastener has an integral flat cap no less than 1" across.

Note: Before starting application of the J-M Flexstone Cap Sheet cut into lengths of 12' to 18' and allow to flatten.

BUILT-UP ROOFS by Johns-Manville / P.O. Box 5108, Denver, Colorado 80217

Regions 1, 2, & 3

*Specification **No. 403-I** for use over*

FESCO, FESCO-FOAM OR OTHER APPROVED INSULATION

*on inclines of **1/4"** to **6"** per foot*

Johns-Manville

Mineral-Surface BUILT-UP ROOFS

This specification is for use over Fesco, Fesco-Foam or any type of approved insulation which is not nailable and which offers suitable surface to receive the roof. This specification is not to be used over light-weight insulating concrete decks either poured or pre-cast, or over fill made of lightweight insulating concrete.

Note: All information contained in "General Instructions" in the current Specification Manual for Johns-Manville Built-Up Roofs shall be considered part of this specification.

Flashings

See section on FLASHINGS, Specification Manual for J-M Built-Up Roofs.

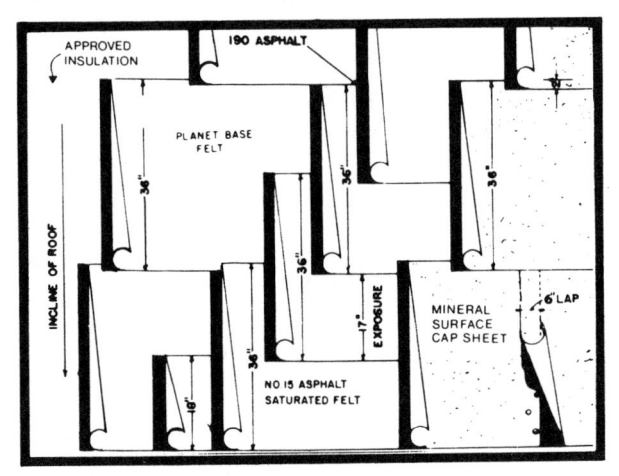

On slopes up to 2" apply felts perpendicular to the slope starting at the low point of each slope. On slopes over 2" apply felts parallel to the slope, nailing at the top of each run of felt on not over 9" centers. If run of felt exceeds 20' an additional line of nails shall be used at 20' intervals.

This specification is eligible for a 15-Year Guarantee only when, in the opinion of an authorized J-M Representative, all conditions listed in "General Instructions" of this Specification Manual have been met.

Application of Roofing

First: Apply one layer of J-M Planet Base Felt, lapping each felt 4" over the preceding one. Mop the full width under each felt with asphalt at a minimum rate of 33 lbs. per square.

Second: Apply two layers of J-M No. 15 Asphalt Saturated Felt (perforated), lapping each felt 19" over the preceding one. Mop the full width of each felt with asphalt at a minimum rate of 23 lbs. per square. Broom each felt so that it shall be firmly and uniformly set without voids into the hot asphalt. On slopes 2" per ft. or greater nail each felt at approximately 9" centers adjacent to the back edge.

Third: Starting at the low edge (on slopes up to 2") or at the side opposite the prevailing wind (on slopes over 2"), apply one layer of J-M Cap Sheet, lapping each layer 2" over the edge of the preceding one. Lap the end 6" over the preceding felt. Mop the full width under each layer with the asphalt, making sure that all edges are well sealed, and with the J-M Cap Sheet uniformly set without voids into the hot asphalt. Back-nail if required.

Fourth: In areas of standing water over the Cap Sheet with slopes of $\frac{1}{4}"$ or less, apply hot asphalt at a rate of about 50 lbs. per square and embed therein approximately 300 lbs. of gravel or slag.

Materials per 100 sq. ft. of roof area

FELTS:
- J-M Planet Base Felt .... 1 layer
- J-M Asphalt Saturated Felt .... 2 layers

REGIONS 1 & 2
J-M GlasKap Mineral Surface Fiber Glass
   Cap Sheet .... 1 layer

REGION 3
J-M Flexstone Mineral Surface Cap Sheet or
J-M GlasKap Mineral Surface Fiber
   Glass Cap Sheet .... 1 layer

BITUMEN:
- J-M 190 Asphalt (Slopes under 3" per foot) ... 102 lbs.*
- J-M 220 Asphalt (Slopes 3" or greater) .... 102 lbs.*

Approximate applied Weight Min: 247 lbs. Max: 252 lbs.

*Deduct 10 lbs. over Fesco-Foam.

Nailing

Where nailing is required, nailing strips must be provided. All nails or other fasteners are to be driven through tin caps unless the nail or fastener has an integral flat cap no less than 1" across.

Note: Before starting application of the J-M Flexstone Cap Sheet cut into lengths of 12' to 18' and allow to flatten.

BUILT-UP ROOFS by Johns-Manville / P.O. Box 5108, Denver, Colorado 80217

Region 3 Only

Specification **No. 404** *for use over*

PLYWOOD OR OTHER NAILABLE DECKS

*on inclines of **1/4"** to **6"** per foot*

Johns-Manville
Mineral-Surface BUILT-UP ROOFS

This specification is to be used over any type of structural deck (without insulation) which can receive and adequately retain nails or other types of mechanical fasteners as may be recommended by the deck manufacturer. Examples of such decks are wood, plywood, and some lightweight aggregate concrete decks* no lighter than 32 PCF density, or that provide proper nail retention.

*Ventsulation Felt must be the base felt over these decks.

Note: All information contained in "General Instructions" in the current Specification Manual for Johns-Manville Built-Up Roofs shall be considered part of this specification.

Flashings
See section on FLASHINGS, Specification Manual for J-M Built-Up Roofs.

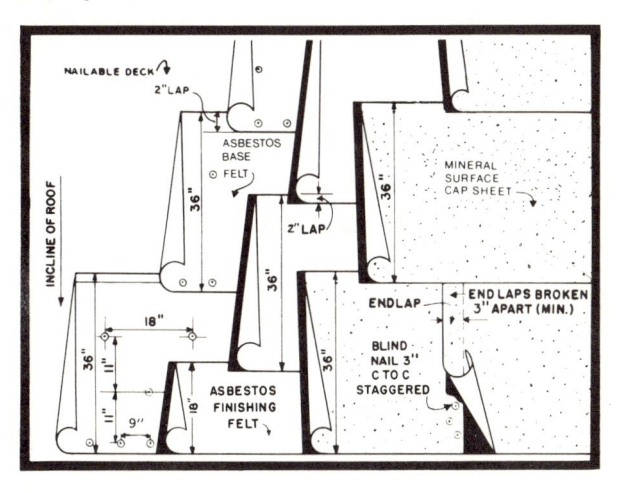

This specification is eligible for a 20-Year Guarantee only when, in the opinion of an authorized J-M Representative, all conditions listed in "General Instructions" of this Specification Manual have been met.

Application of Roofing

Over wood board decks one ply of sheathing paper must be used under the base felt next to the deck.

Over plywood, omit sheathing paper and sprinkle mop base felt using 10 lbs. of asphalt per square.

First: Apply one layer of J-M Asbestos Base Felt starting at the low edge either parallel or perpendicular to the slope, and lapping each course 2" over the preceding one. Nail the laps at 9" centers and down the longitudinal center of each felt nail two rows of nails with the rows spaced approximately 11" apart and nails staggered on approximately 18" centers. Use nails or fasteners appropriate to the type of deck.

Second: Apply one 18" wide layer of J-M Asbestos Finishing Felt (perforated), then continue with full 36" widths, lapping each course 2" over the preceding one. Mop the full width under each felt with asphalt at a minimum rate of 23 lbs. per square.

Third: Starting at the low edge apply one layer of J-M Cap Sheet lapping each layer 2" over the edge of the preceding one. Lap the end 6" over the preceding felt. Mop the full width under each layer with the asphalt at a minimum rate of 23 lbs. per square making sure that all edges are well sealed, and with the Cap Sheet uniformly set without voids into the hot asphalt. On slopes 2" per foot and greater nail as required. Nail laps on 3" centers and mop the full width with asphalt.

Fourth: In areas of standing water over the Cap Sheet with slopes of $1/4$" or less, apply hot asphalt at a rate of about 50 lbs. per square and embed therein approximately 300 lbs. of gravel or slag.

Materials per 100 sq. ft. of roof area

FELTS: J-M Centurian or Coated Asbestos Base Felt ..... 1 layer
J-M Asbestos Finishing Felt
(Perforated) .... 1 layer
J-M Flexstone Mineral Surface Cap Sheet or
J-M GlasKap Mineral Surface Fiber
Glass Cap Sheet .... 1 layer

BITUMEN:
J-M 190 Asphalt (Slopes under 3" per foot) ..... 46 lbs.
J-M 220 Asphalt (Slopes 3" or greater) .... 46 lbs.
Sprinkle mopping to plywood .... 10 lbs.

Approximate applied Weight Min: 176 lbs. Max: 191 lbs.

Nailing

All nails or other fasteners are to be driven through tin caps unless the nail or fastener has an integral flat cap no less than 1" across.

Note: Before starting application of the J-M Flexstone Cap Sheet cut into lengths of 12' to 18' and allow to flatten.

BUILT-UP ROOFS by Johns-Manville / P.O. Box 5108, Denver, Colorado 80217

Region 3 Only

Specification **No. 405** *for use over*

CONCRETE OR OTHER NON-NAILABLE DECKS

*on inclines of **1/4"** to **6"** per foot*

Johns-Manville
Mineral-Surface BUILT-UP ROOFS

This specification is for use over any type of structural deck which is not nailable and which offers suitable surface to receive the roof. Poured and pre-cast concrete decks require priming. This specification is not to be used over lightweight insulating concrete decks either poured or pre-cast, or over fill made of lightweight insulating concrete.

Note: All information contained in "General Instructions" in the current Specification Manual for Johns-Manville Built-Up Roofs shall be considered part of this specification.

Flashings

See section on FLASHINGS, Specification Manual for J-M Built-Up Roofs.

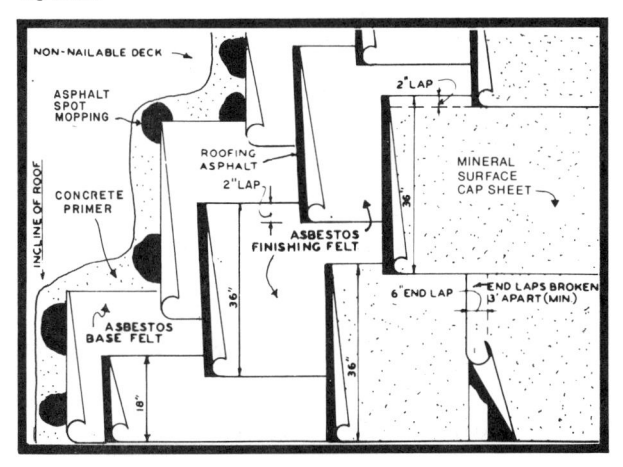

On slopes up to 2" apply felts perpendicular to the slope starting at the low point of each slope. On slopes over 2" apply felts parallel to the slope, nailing at the top of each run of felt on not over 9" centers. If run of felt exceeds 20' an additional line of nails shall be used at 20' intervals.

This specification is eligible for a 20-Year Guarantee only when, in the opinion of an authorized J-M Representative, all conditions listed in "General Instructions" of this Specification Manual have been met.

Application of Roofing

First: Apply one layer J-M Asbestos Base Felt lapping each felt two inches over the preceding one. Spot mop the full width under each felt with 190-Asphalt using a minimum of 10 lbs. per square.

Second: Apply one 18" wide layer of J-M Asbestos Finishing Felt, then continue with full 36" widths, lapping each course 2" over the preceding one. Mop the full width under each felt with asphalt using at a minimum rate of 23 lbs. per square. Nail as required.

Third: Starting at the low edge (on slopes up to 2") or at the side opposite the prevailing wind (on slopes over 2"), apply one layer of J-M Cap Sheet, lapping each layer 2" over the preceding one. Lap the end 6" over the preceding felt. Mop the full width under each layer with the asphalt at a minimum rate of 23 lbs. per square making sure that all edges are well sealed, and with the J-M Cap Sheet uniformly set without voids into the hot asphalt. Back-nail if required.

Fourth: In areas of standing water over the Cap Sheet with slopes of 1/4" or less, apply hot asphalt at a rate of about 50 lbs. per square and embed therein approximately 300 lbs. of gravel or slag.

Materials per 100 sq. ft. of roof area

CONCRETE
PRIMER:
        If required .... 1 gal.

FELTS: J-M Centurian or Coated Asbestos Base Felt ..... 1 layer
          J-M Asbestos Finishing Felt .... 1 layer
          J-M Flexstone Mineral Surface Cap Sheet or
          J-M GlasKap Mineral Surface Fiber
             Glass Cap Sheet .... 1 layer

BITUMEN:
        J-M 190 Asphalt (Slopes under 3" per foot) ..... 56 lbs.
        J-M 220 Asphalt (Slopes 3" or greater) .... 56 lbs.

Approximate applied Weight Min: 168 lbs. Max: 191 lbs.

Nailing

Where nailing is required, nailing strips must be provided. All nails or other fasteners are to be driven through tin caps unless the nail or fastener has an integral flat cap no less than 1" across.

Note: Before starting application of the J-M Flexstone Cap Sheet cut into lengths of 12' to 18' and allow to flatten.

BUILT-UP ROOFS by Johns-Manville / P.O. Box 5108, Denver, Colorado 80217

Region 3 Only

Specification **No. 405-1** *for use over*

FESCO, FESCO-FOAM OR OTHER APPROVED INSULATION

*on inclines of **1/4"** to **6"** per foot*

Johns-Manville
Mineral-Surface BUILT-UP ROOFS

This specification is for use over Fesco, Fesco-Foam or any type of approved insulation which is not nailable and which offers suitable surface to receive the roof. This specification is not to be used over light-weight insulating concrete decks either poured or pre-cast, or over fill made of lightweight insulating concrete.

Note: All information contained in "General Instructions" in the current Specification Manual for Johns-Manville Built-Up Roofs shall be considered part of this specification.

Flashings

See section on FLASHINGS, Specification Manual for J-M Built-Up Roofs.

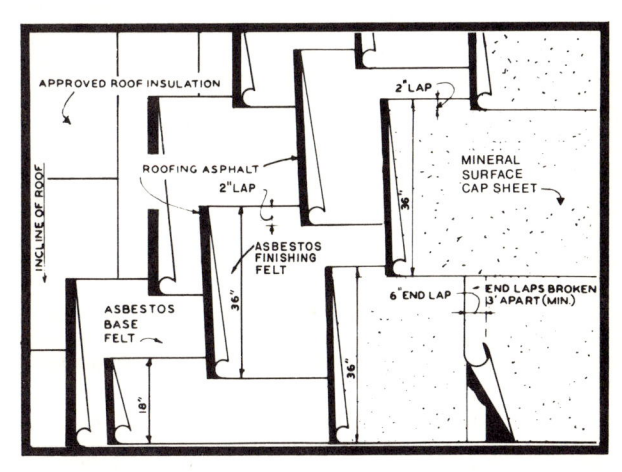

On slopes up to 2" apply felts perpendicular to the slope starting at the low point of each slope. On slopes over 2" apply felts parallel to the slope, nailing at the top of each run of felt on not over 9" centers. If run of felt exceeds 20' an additional line of nails shall be used at 20' intervals.

This specification is eligible for a 20-Year Guarantee only when, in the opinion of an authorized J-M Representative, all conditions listed in "General Instructions" of this Specification Manual have been met.

Application of Roofing

First: Apply one layer J-M Asbestos Base Felt lapping each felt two inches over the preceding one. Mop the full width under each felt with 190-Asphalt using a minimum of 33 lbs. per square.

Second: Apply one 18" wide layer of J-M Asbestos Finishing Felt, then continue with full 36" widths, lapping each course 2" over the preceding one. Mop the full width under each felt with asphalt using at a minimum rate of 23 lbs. per square. Nail as required.

Third: Starting at the low edge (on slopes up to 2") or at the side opposite the prevailing wind (on slopes over 2"), apply one layer of J-M Cap Sheet, lapping each layer 2" over the preceding one. Lap the end 6" over the preceding felt. Mop the full width under each layer with the asphalt at a minimum rate of 23 lbs. per square making sure that all edges are well sealed, and with the J-M Cap Sheet uniformly set without voids into the hot asphalt. Back-nail if required.

Fourth: In areas of standing water over the Cap Sheet with slopes of $\frac{1}{4}$" or less, apply hot asphalt at a rate of about 50 lbs. per square and embed therein approximately 300 lbs. of gravel or slag.

Materials per 100 sq. ft. of roof area

FELTS:
- J-M Centurian or Coated Asbestos Base Felt ..... 1 layer
- J-M Asbestos Finishing Felt .... 1 layer
- J-M Flexstone Mineral Surface Cap Sheet or
- J-M GlasKap Mineral Surface Fiber Glass Cap Sheet .... 1 layer

BITUMEN:
- J-M 190 Asphalt (Slopes under 3" per foot) ..... 79 lbs.*
- J-M 220 Asphalt (Slopes 3" or greater) .... 79 lbs.*

Approximate applied Weight Min: 209 lbs. Max: 214 lbs.

*Deduct 10 lbs. over Fesco-Foam.

Nailing

Where nailing is required, nailing strips must be provided. All nails or other fasteners are to be driven through tin caps unless the nail or fastener has an integral flat cap no less than 1" across.

Note: Before starting application of the J-M Flexstone Cap Sheet cut into lengths of 12' to 18' and allow to flatten.

BUILT-UP ROOFS by Johns-Manville / P.O. Box 5108, Denver, Colorado 80217

Region 3 Only

Specification **No. 406** *for use over*

PLYWOOD DECKS

on inclines of 1/4" to 6" per foot

Johns-Manville
Mineral-Surface BUILT-UP ROOFS

This specification is for use over plywood structural deck (without insulation) which can receive and adequately retain nails or other types of mechanical fasteners as may be recommended by the deck manufacturer.

Note: All information contained in "General Instructions" in the current Specification Manual for Johns-Manville Built-Up Roofs shall be considered part of this specification.

Flashings

See section on FLASHINGS, Specification Manual for J-M Built-Up Roofs.

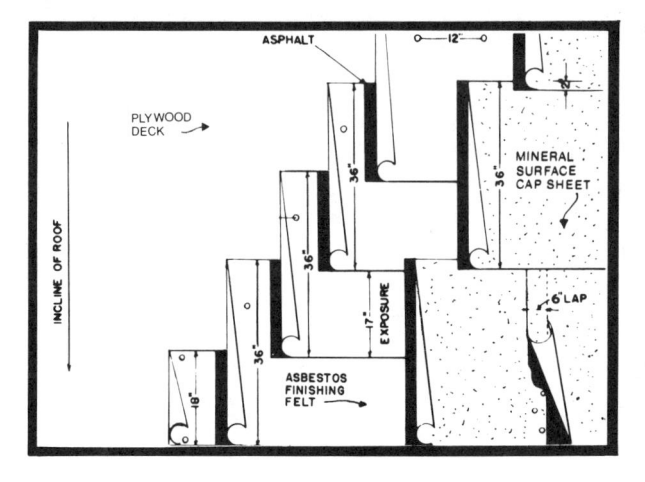

This specification is eligible for a 20-Year Guarantee only when, in the opinion of an authorized J-M Representative, all conditions listed in "General Instructions" of this Specification Manual have been met.

Application of Roofing

Over plywood, sprinkle mop to the deck and solidly mop under the overlap, omit nailing if slope allows.

First: Apply two layers of J-M Asbestos Finishing Felt (perforated), lapping each felt 19" over the preceding one. Mop only the width under the overlap of each felt with asphalt at a minimum rate of 23 lbs. per square, and nail on 12" centers, 9" down from the back edge. Broom each felt so that it shall be firmly and uniformly set without voids into the hot asphalt.

Second: Starting at the low edge (on slopes up to 2") or at the side opposite the prevailing wind (on slopes over 2"), apply one layer of J-M Cap Sheet, lapping each layer 2" over the preceding one. Lap the end 6" over the preceding felt. Mop the full width under each layer with the asphalt at a minimum rate of 23 lbs. per square making sure that all edges are well sealed, and with the J-M Cap Sheet uniformly set without voids into the hot asphalt. Back-nail if required.

Third: In areas of standing water over the Cap Sheet for slopes of $\frac{1}{4}$" or less, apply hot asphalt at a rate of about 50 lbs. per square and embed therein approximately 300 lbs. of gravel or slag.

Materials per 100 sq. ft. of roof area

FELTS: J-M Asbestos Finishing Felt .... 2 layers
J-M Flexstone Mineral Surface Cap Sheet or
J-M GlasKap Mineral Surface Fiber
Glass Cap Sheet .... 1 layer

BITUMEN:
J-M 190 Asphalt (Slopes under 3" per foot) .... 46 lbs.
J-M 220 Asphalt (Slopes 3" or greater) .... 46 lbs.
Sprinkle mopping to plywood .... 10 lbs.

Approximate applied Weight Min: 148 lbs. Max: 163 lbs.

Nailing

All nails or other fasteners are to be driven through tin caps unless the nail or fastener has an integral flat cap no less than 1" across.

Note: Before starting application of the J-M Flexstone Cap Sheet cut into lengths of 12' to 18' and allow to flatten.

BUILT-UP ROOFS *by Johns-Manville / P.O. Box 5108, Denver, Colorado 80217*

Region 3 Only

*Specification **No. 407** for use over*

CONCRETE OR OTHER NON-NAILABLE DECKS

*on inclines of **1/4"** to **6"** per foot*

Johns-Manville
Mineral-Surface BUILT-UP ROOFS

This specification is for use over any type of structural deck which is not nailable and which offers suitable surface to receive the roof. Poured and pre-cast concrete decks require priming. This specification is not to be used over lightweight insulating concrete decks either poured or pre-cast, or over fill made of lightweight insulating concrete.

Note: All information contained in "General Instructions" in the current Specification Manual for Johns-Manville Built-Up Roofs shall be considered part of this specification.

Flashings

See section on FLASHINGS, Specification Manual for J-M Built-Up Roofs.

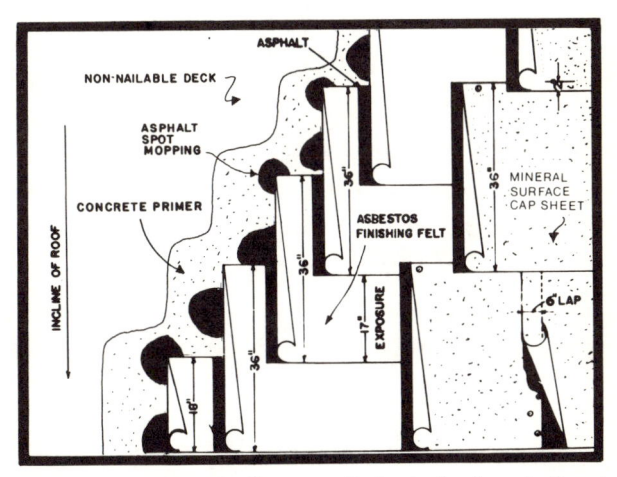

On slopes up to 2" apply felts perpendicular to the slope starting at the low point of each slope. On slopes over 2" apply felts parallel to the slope, nailing at the top of each run of felt on not over 9" centers. If run of felt exceeds 20' an additional line of nails shall be used at 20' intervals.

This specification is eligible for a 20-Year Guarantee only when, in the opinion of an authorized J-M Representative, all conditions listed in "General Instructions" of this Specification Manual have been met.

Application of Roofing

First: Apply two layers of J-M Asbestos Finishing Felt (perforated), lapping each felt 19" over the preceding one. Mop only the width under the overlap of each felt with asphalt at a minimum rate of 23 lbs. per square, and spot mop to the deck using 10 lbs. of asphalt per square. Broom each felt so that it shall be firmly and uniformly set without voids into the hot asphalt. Nail as required.

Second: Starting at the low edge (on slopes up to 2") or at the side opposite the prevailing wind (on slopes over 2"), apply one layer of J-M Cap Sheet, lapping each layer 2" over the preceding one. Lap the end 6" over the preceding felt. Mop the full width under each layer with the asphalt at a minimum rate of 23 lbs. per square making sure that all edges are well sealed, and with the J-M Cap Sheet uniformly set without voids into the hot asphalt. Back-nail if required.

Third: In areas of standing water over the Cap Sheet for slopes of ¼" or less, apply hot asphalt at a rate of about 50 lbs. per square and embed therein approximately 300 lbs. of gravel or slag.

Materials per 100 sq. ft. of roof area

CONCRETE
PRIMER:
        If required .... 1 gal.

FELTS: J-M Asbestos Finishing Felt .... 2 layers
            J-M Flexstone Mineral Surface Cap Sheet or
            J-M GlasKap Mineral Surface Fiber
                Glass Cap Sheet .... 1 layer

BITUMEN:
        J-M 190 Asphalt (Slopes under 3" per foot) ..... 56 lbs.
        J-M 220 Asphalt (Slopes 3" or greater) .... 56 lbs.

Approximate applied Weight Min: 158 lbs. Max: 163 lbs.

Nailing

Where nailing is required, nailing strips must be provided. All nails or other fasteners are to be driven through tin caps unless the nail or fastener has an integral flat cap no less than 1" across.

Note: Before starting application of the J-M Flexstone Cap Sheet cut into lengths of 12' to 18' and allow to flatten.

BUILT-UP ROOFS by Johns-Manville / P.O. Box 5108, Denver, Colorado 80217

Specification **No. 407-I** *for use over*

FESCO, FESCO-FOAM OR OTHER APPROVED INSULATION

on inclines of 1/4" to 6" per foot

Johns-Manville
Mineral-Surface BUILT-UP ROOFS

This specification is for use over Fesco, Fesco-Foam or any type of approved insulation which is not nailable and which offers suitable surface to receive the roof. This specification is not to be used over light-weight insulating concrete decks either poured or pre-cast, or over fill made of lightweight insulating concrete.

Note: All information contained in "General Instructions" in the current Specification Manual for Johns-Manville Built-Up Roofs shall be considered part of this specification.

Flashings

See section on FLASHINGS, Specification Manual for J-M Built-Up Roofs.

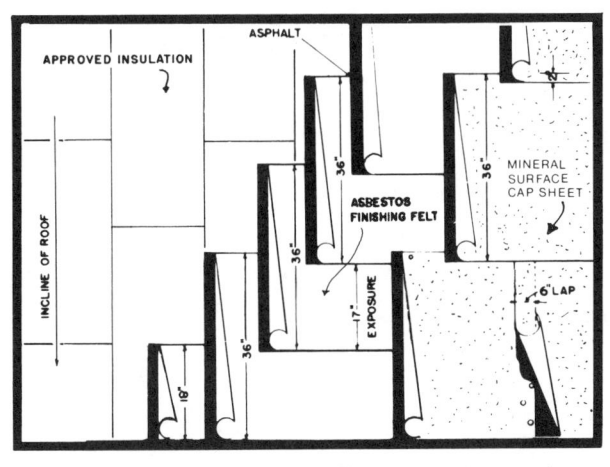

On slopes up to 2" apply felts perpendicular to the slope starting at the low point of each slope. On slopes over 2" apply felts parallel to the slope, nailing at the top of each run of felt on not over 9" centers. If run of felt exceeds 20' an additional line of nails shall be used at 20' intervals.

This specification is eligible for a 20-Year Guarantee only when, in the opinion of an authorized J-M Representative, all conditions listed in "General Instructions" of this Specification Manual have been met.

Application of Roofing

First: Apply two layers of J-M Asbestos Finishing Felt (perforated), lapping each felt 19" over the preceding one. Mop the full width under each felt with asphalt at a minimum rate of 33 lbs. per square. Broom each felt so that it shall be firmly and uniformly set without voids into the hot asphalt. Nail as required.

Second: Starting at the low edge (on slopes up to 2") or at the side opposite the prevailing wind (on slopes over 2"), apply one layer of J-M Cap Sheet, lapping each layer 2" over the preceding one. Lap the end 6" over the preceding felt. Mop the full width under each layer with the asphalt at a minimum rate of 23 lbs. per square making sure that all edges are well sealed, and with the J-M Cap Sheet uniformly set without voids into the hot asphalt. Back-nail if required.

Third: In areas of standing water over the Cap Sheet for slopes of ¼" or less, apply hot asphalt at a rate of about 50 lbs. per square and embed therein approximately 300 lbs. of gravel or slag.

Materials per 100 sq. ft. of roof area

FELTS: J-M Asbestos Finishing Felt .... 2 layers
J-M Flexstone Mineral Surface Cap Sheet or
J-M GlasKap Mineral Surface Fiber
Glass Cap Sheet .... 1 layer

BITUMEN:
J-M 190 Asphalt (Slopes under 3" per foot) ..... 79 lbs.*
J-M 220 Asphalt (Slopes 3" or greater) .... 79 lbs.*

Approximate applied Weight Min: 184 lbs. Max: 189 lbs.

*Deduct 10 lbs. over Fesco-Foam.

Nailing

Where nailing is required, nailing strips must be provided. All nails or other fasteners are to be driven through tin caps unless the nail or fastener has an integral flat cap no less than 1" across.

Note: Before starting application of the J-M Flexstone Cap Sheet cut into lengths of 12' to 18' and allow to flatten.

BUILT-UP ROOFS by Johns-Manville / P.O. Box 5108, Denver, Colorado 80217

Region 3 Only

*Specification **No. 408** for use over*

PLYWOOD DECKS

*on inclines of **1/4"** to **6"** per foot*

Johns-Manville
Mineral-Surface BUILT-UP ROOFS

This specification is for use over plywood structural deck (without insulation) which can receive and adequately retain nails or other types of mechanical fasteners as may be recommended by the deck manufacturer.

Note: All information contained in "General Instructions" in the current Specification Manual for Johns-Manville Built-Up Roofs shall be considered part of this specification.

Flashings

See section on FLASHINGS, Specification Manual for J-M Built-Up Roofs.

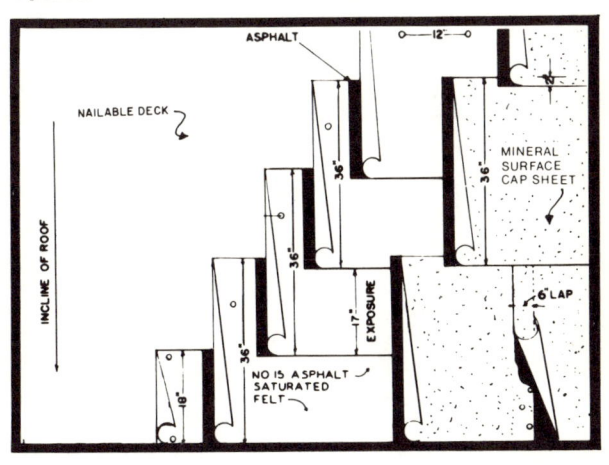

This specification is eligible for a 15-Year Guarantee only when, in the opinion of an authorized J-M Representative, all conditions listed in "General Instructions" of this Specification Manual have been met.

Application of Roofing

Over plywood, sprinkle mop to the deck and solidly mop under the overlap, omit nailing if slope allows.

First: Apply two layers of J-M Asphalt Saturated Felt (perforated), lapping each felt 19" over the preceding one. Mop only the width under the overlap of each felt with asphalt at a minimum rate of 23 lbs. per square, and nail on 12" centers, 9" down from the back edge. Broom each felt so that it shall be firmly and uniformly set without voids into the hot asphalt.

Second: Starting at the low edge (on slopes up to 2") or at the side opposite the prevailing wind (on slopes over 2"), apply one layer of J-M Cap Sheet, lapping each layer 2" over the preceding one. Lap the end 6" over the preceding felt. Mop the full width under each layer with the asphalt at a minimum rate of 23 lbs. per square making sure that all edges are well sealed, and with the J-M Cap Sheet uniformly set without voids into the hot asphalt. Back-nail if required.

Third: In areas of standing water over the Cap Sheet for slopes of ¼" or less, apply hot asphalt at a rate of about 50 lbs. per square and embed therein approximately 300 lbs. of gravel or slag.

Materials per 100 sq. ft. of roof area

FELTS:	J-M Asphalt Saturated Felt	2 layers
	J-M Flexstone Mineral Surface Cap Sheet or J-M GlasKap Mineral Surface Fiber Glass Cap Sheet	1 layer
BITUMEN:		
	J-M 190 Asphalt (Slopes under 3" per foot)	46 lbs.
	J-M 220 Asphalt (Slopes 3" or greater)	46 lbs.
	Sprinkle mopping to plywood	10 lbs.

Approximate applied Weight Min: 148 lbs. Max: 163 lbs.

Nailing

All nails or other fasteners are to be driven through tin caps unless the nail or fastener has an integral flat cap no less than 1" across.

Note: Before starting application of the J-M Flexstone Cap Sheet cut into lengths of 12' to 18' and allow to flatten.

BUILT-UP ROOFS by Johns-Manville / P.O. Box 5108, Denver, Colorado 80217

or Region 3 Only

Specification **No. 409** *for use over*

CONCRETE OR OTHER NON-NAILABLE DECKS

on inclines of ***1/4"*** *to* ***6"*** *per foot*

Johns-Manville
Mineral-Surface BUILT-UP ROOFS

This specification is for use over any type of structural deck which is not nailable and which offers suitable surface to receive the roof. Poured and pre-cast concrete decks require priming. This specification is not to be used over lightweight insulating concrete decks either poured or pre-cast, or over fill made of lightweight insulating concrete.

Note: All information contained in "General Instructions" in the current Specification Manual for Johns-Manville Built-Up Roofs shall be considered part of this specification.

Flashings

See section on FLASHINGS, Specification Manual for J-M Built-Up Roofs.

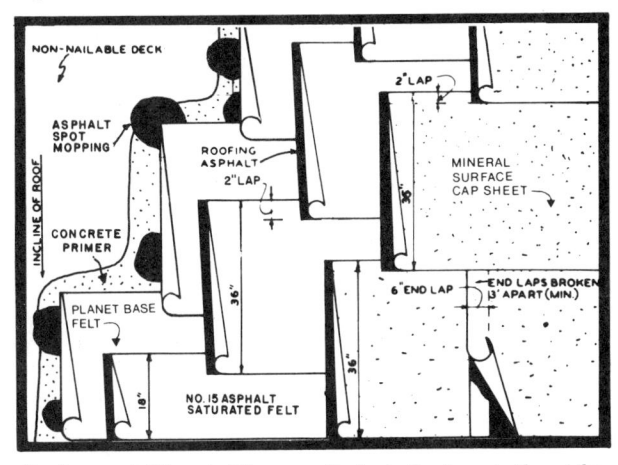

On slopes up to 2" apply felts perpendicular to the slope starting at the low point of each slope. On slopes over 2" apply felts parallel to the slope, nailing at the top of each run of felt on not over 9" centers. If run of felt exceeds 20' an additional line of nails shall be used at 20' intervals.

This specification is eligible for a 15-Year Guarantee only when, in the opinion of an authorized J-M Representative, all conditions listed in "General Instructions" of this Specification Manual have been met.

Application of Roofing

First: Apply one layer of J-M Planet Base Felt lapping each felt 4" over the preceding one. Spot mop the full width under each felt with asphalt using a minimum of 10 lbs. per square.

Second: Apply one 18" wide layer of J-M Asphalt Saturated Felt, then continue with full 36" widths, lapping each course 2" over the preceding one. Mop the full width under each felt with asphalt using at a minimum rate of 23 lbs. per square. Nail as required.

Third: Starting at the low edge (on slopes up to 2") or at the side opposite the prevailing wind (on slopes over 2"), apply one layer of J-M Cap Sheet, lapping each layer 2" over the preceding one. Lap the end 6" over the preceding felt. Mop the full width under each layer with the asphalt at a minimum rate of 23 lbs. per square making sure that all edges are well sealed, and with the J-M Cap Sheet uniformly set without voids into the hot asphalt. Back-nail if required.

Fourth: In areas of standing water over the Cap Sheet with slopes of 1/4" or less, apply hot asphalt at a rate of about 50 lbs. per square and embed therein approximately 300 lbs. of gravel or slag.

Materials per 100 sq. ft. of roof area

CONCRETE
PRIMER:
        If required .... 1 gal.

FELTS: J-M Planet Base Felt .... 1 layer
            J-M Asphalt Saturated Felt .... 1 layer
            J-M Flexstone Mineral Surface Cap Sheet or
            J-M GlasKap Mineral Surface Fiber
                Glass Cap Sheet .... 1 layer

BITUMEN:
        J-M 190 Asphalt (Slopes under 3" per foot) .... 56 lbs.
        J-M 220 Asphalt (Slopes 3" or greater) .... 56 lbs.

Approximate applied Weight Min: 186 lbs. Max: 191 lbs.

Nailing

Where nailing is required, nailing strips must be provided. All nails or other fasteners are to be driven through tin caps unless the nail or fastener has an integral flat cap no less than 1" across.

Note: Before starting application of the J-M Flexstone Cap Sheet cut into lengths of 12' to 18' and allow to flatten.

BUILT-UP ROOFS by Johns-Manville / P.O. Box 5108, Denver, Colorado 80217

Region 3 Only

*Specification **No. 409-I** for use over*

FESCO, FESCO-FOAM OR OTHER APPROVED INSULATION

*on inclines of **1/4"** to **6"** per foot*

Johns-Manville
Mineral-Surface BUILT-UP ROOFS

This specification is for use over Fesco, Fesco-Foam or any type of approved insulation which is not nailable and which offers suitable surface to receive the roof. This specification is not to be used over light-weight insulating concrete decks either poured or pre-cast, or over fill made of lightweight insulating concrete.

Note: All information contained in "General Instructions" in the current Specification Manual for Johns-Manville Built-Up Roofs shall be considered part of this specification.

Flashings

See section on FLASHINGS, Specification Manual for J-M Built-Up Roofs.

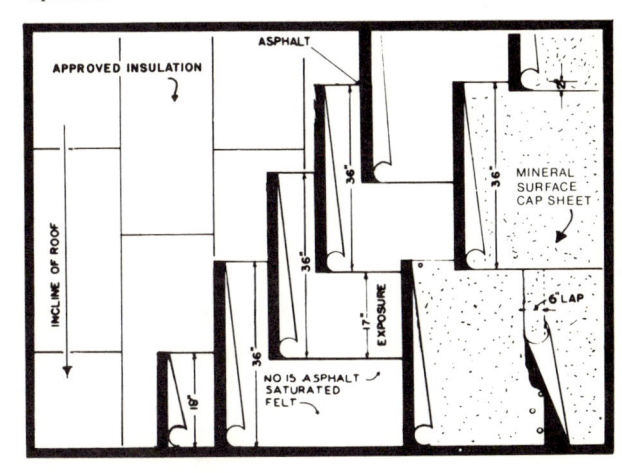

On slopes up to 2" apply felts perpendicular to the slope starting at the low point of each slope. On slopes over 2" apply felts parallel to the slope, nailing at the top of each run of felt on not over 9" centers. If run of felt exceeds 20' an additional line of nails shall be used at 20' intervals.

This specification is eligible for a 15-Year Guarantee only when, in the opinion of an authorized J-M Representative, all conditions listed in "General Instructions" of this Specification Manual have been met.

Application of Roofing

First: Apply two layers of J-M Asphalt Saturated Felt (perforated), lapping each felt 19" over the preceding one. Mop the full width under each felt with asphalt at a minimum rate of 33 lbs. per square. Broom each felt so that it shall be firmly and uniformly set without voids into the hot asphalt. Nail as required.

Second: Starting at the low edge (on slopes up to 2") or at the side opposite the prevailing wind (on slopes over 2"), apply one layer of J-M Cap Sheet, lapping each layer 2" over the preceding one. Lap the end 6" over the preceding felt. Mop the full width under each layer with the asphalt at a minimum rate of 23 lbs. per square making sure that all edges are well sealed, and with the J-M Cap Sheet uniformly set without voids into the hot asphalt. Back-nail if required.

Third: In areas of standing water over the Cap Sheet for slopes of ¼" or less, apply hot asphalt at a rate of about 50 lbs. per square and embed therein approximately 300 lbs. of gravel or slag.

Materials per 100 sq. ft. of roof area

FELTS: J-M Asphalt Saturated Felt .... 2 layers
J-M Flexstone Mineral Surface Cap Sheet or
J-M GlasKap Mineral Surface Fiber
   Glass Cap Sheet .... 1 layer

BITUMEN:
  J-M 190 Asphalt (Slopes under 3" per foot) ..... 79 lbs.*
  J-M 220 Asphalt (Slopes 3" or greater) .... 79 lbs.*

Approximate applied Weight Min: 184 lbs. Max: 189 lbs.

*Deduct 10 lbs. over Fesco-Foam.

Nailing

Where nailing is required, nailing strips must be provided. All nails or other fasteners are to be driven through tin caps unless the nail or fastener has an integral flat cap no less than 1" across.

Note: Before starting application of the J-M Flexstone Cap Sheet cut into lengths of 12' to 18' and allow to flatten.

BUILT-UP ROOFS by Johns-Manville/P.O. Box 5108, Denver, Colorado 80217

Specification **No. 420 VS-1** *for use over*

ANY ACCEPTABLE DECK

on inclines of 1/4" to 6" per foot

Johns-Manville
Mineral-Surface BUILT-UP ROOFS

This specification is for use over any approved structural deck or insulation. Over nailable decks, all felts are to be back-nailed as described below. On inclines over 2" per foot, felts must be nailed into nailing strips or into insulation with appropriate fasteners to prevent felt slippage.

Note: All information contained in "General Instructions" in the current Specification Manual for Johns-Manville Built-Up Roofs shall be considered part of this specification.

Flashings
See section on FLASHINGS, Specification Manual for J-M Built-Up Roofs.

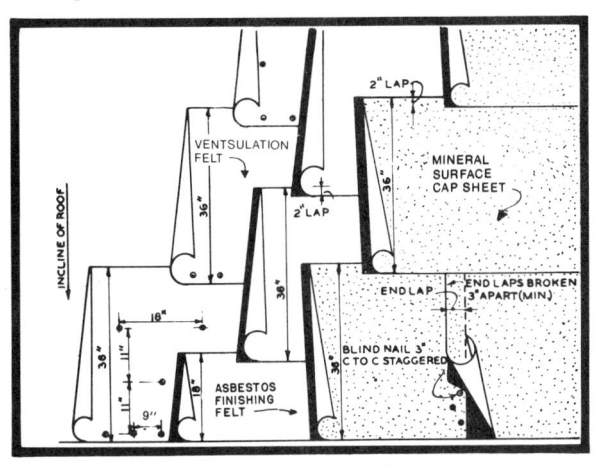

This specification is eligible for a 20-Year Guarantee only when, in the opinion of an authorized J-M Representative, all conditions listed in "General Instructions" of this Specification Manual have been met.

Application of Roofing

On slopes up to 2" apply felts perpendicular to the slope starting at the low point of each slope. On slopes over 2" apply felts parallel to the slope, nailing at the top of each run of felt on not over 9" centers. If run of felt exceeds 20' an additional line of nails shall be used at 20' intervals.

First: Apply one layer of J-M Ventsulation Felt starting at the low edge and working up the slope, lapping each felt the 1" selvage over the preceding one and, depending on type of deck, nailing as shown. Ventsulation Felt must be vented at parapets and roof edges in accordance with Specifications VS-1.

Second: Apply one 18" wide layer of J-M Asbestos Finishing Felt (perforated), then continue with full 36" widths, lapping each course 2" over the preceding one. Broom each felt so that it shall be firmly and uniformly set without voids into hot asphalt applied just before the felt at a minimum rate of 23 lbs. per square uniformly over the entire surface.

Third: Starting at the low edge apply one layer of J-M Cap Sheet lapping each layer 2" over the edge of the preceding one. Lap the end 6" over the preceding felt. Mop the full width under each layer with the asphalt at a minimum rate of 23 lbs. per square making sure that all edges are well sealed, and with the Cap Sheet uniformly set without voids into the hot asphalt. On slopes 2" per foot and greater nail as required. Nail laps on 3" centers and mop the full width with asphalt.

Fourth: In areas of standing water over the Cap Sheet with slopes of 1/4" or less, apply hot asphalt at a rate of about 50 lbs. per square and embed therein approximately 300 lbs. of gravel or slag.

Materials per 100 sq. ft. of roof area

CONCRETE
PRIMER:
        If required .... 1 gal.

FELTS: J-M Ventsulation Felt .... 1 layer
              J-M Asbestos Finishing Felt
              (Perforated) .... 1 layer
              J-M Flexstone Mineral Surface Cap Sheet or
              J-M GlasKap Mineral Surface Fiber
                Glass Cap Sheet .... 1 layer

BITUMEN:
        J-M 190 Asphalt (Slopes under 3" per foot) ..... 46 lbs.
        J-M 220 Asphalt (Slopes 3" or greater) .... 46 lbs.

Approximately applied Weight Min: 208 lbs. Max: 223 lbs.

Nailing

Where nailing is required, nailing strips must be provided. All nails or other fasteners are to be driven through tin caps unless the nail or fastener has an integral flat cap no less than 1" across.

Note: Before starting application of the J-M Flexstone Cap Sheet cut into lengths of 12' to 18' and allow to flatten.

BUILT-UP ROOFS by Johns-Manville / P.O. Box 5108, Denver, Colorado 80217

Johns-Manville
Gravel Surface
BUILT-UP ROOFS

GRAVEL SURFACE

Specification **No. 600** *for use over*

PLYWOOD OR OTHER NAILABLE DECKS

on inclines of up to 1/2" per foot

Johns-Manville
Gravel-Surface Asbestos BUILT-UP ROOFS

This specification is to be used over any type of structural deck (without insulation) which can receive and adequately retain nails or other types of mechanical fasteners as may be recommended by the deck manufacturer. Examples of such decks are wood, plywood, and some lightweight aggregate concrete decks* no lighter than 32 PCF density, or that provide proper nail retention.

*Ventsulation Felt must be the base felt over these decks.

Note: All information contained in "General Instructions" in the current Specification Manual for Johns-Manville Built-Up Roofs shall be considered part of this specification.

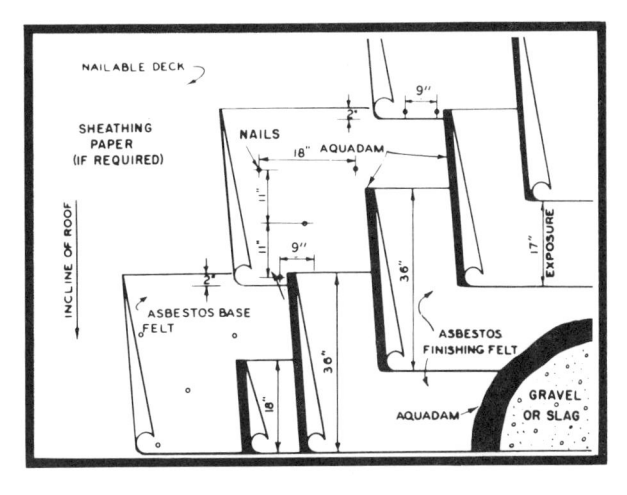

This specification is eligible for a 15-Year Guarantee in Region 1 and 20-Year Guarantee in Regions 2 & 3 only when, in the opinion of an authorized J-M Representative, all conditions listed in "General Instructions" of this Specification Manual have been met.

Application of Roofing

Over wood board decks one ply of sheathing paper must be used under the base felt next to the deck.

Regions 1 & 2: Use J-M Centurian as the base felt.

Region 3: Use either J-M Centurian or Coated Asbestos as the base felt.

First: Start at the low edge and working up the slope and perpendicular to the slope and lapping each felt 2" over the preceding one. Nail the laps at 9" centers and down the longitudinal center of each felt nail two rows of nails with rows spaced approximately 11" apart and nails staggered on approximately 18" centers. Use nails for fasteners appropriate to the type of deck.

Note: In Region 3 only if deck is plywood, base felt may be sprinkle mopped using 10 lbs. of asphalt per square.

Second: Starting at the low edge apply one 18" wide, then over that one 36' wide J-M Asbestos Finishing Felt. Following felts are to be applied full width overlapping the preceding felt by 19" in such manner that at least 2 plies of felt cover the base felt at any point. Broom each felt so that it shall be firmly and uniformly set without voids into hot Aquadam applied just before the felt at a minimum rate of 23 lbs. per square uniformly over the entire surface.

Third: Flood the surface with J-M Aquadam at a minimum rate of 60 lbs per square and while it is still hot embed therein an acceptable gravel at the rate of approximately 400 lbs per square or an acceptable slag at a rate of approximately 300 lbs per square.

Materials per 100 sq ft of roof area

FELTS:
- J-M Centurian or Coated Asbestos Base Felt .... 1 layer
- J-M Asbestos Finishing Felt .... 2 layers

BITUMEN:
- J-M Aquadam .... 46 lbs
- Sprinkle mopping to plywood .... 10 lbs

SURFACING:
- J-M Aquadam .... 60 lbs
- Gravel .... 400 lbs
- Slag .... 300 lbs

Approximately applied Weight Min: 461 lbs. Max: 589 lbs.

Nailing

All nails or other fasteners are to be driven through tin caps unless the nail or fastener has an integral flat cap no less than 1" across.

BUILT-UP ROOFS by Johns-Manville/Greenwood Plaza, Denver, Colorado 80217

Regions 1, 2, & 3

Specification No. 601 for use over

CONCRETE OR OTHER NON-NAILABLE DECKS

on inclines of up to 1/2" per foot

Johns-Manville

Gravel-Surface Asbestos BUILT-UP ROOFS

This specification is for use over any type of structural deck which is not nailable and which offers suitable surface to receive the roof. Poured and pre-cast concrete decks require priming. This specification is not to be used over lightweight insulating concrete decks either poured or pre-cast, or over fill made of lightweight insulating concrete.

Note: All information contained in "General Instructions" in the current Specification Manual for Johns-Manville Built-Up Roofs shall be considered part of this specification.

Flashings

See section on FLASHINGS, Specification Manual for J-M Built-Up Roofs.

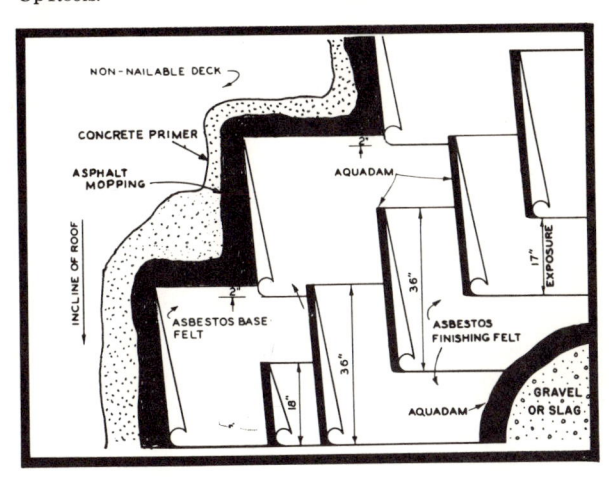

This specification is eligible for a 20-Year Guarantee only when, in the opinion of an authorized J-M Representative, all conditions listed in "General Instructions" of this Specification Manual have been met.

Application of Roofing

First: Regions 1 & 2: Use J-M Centurian as the base felt, lapping each felt 2" over the preceding one and solidly mop the full width under each ply felt with asphalt using a minimum of 23 lbs. per square.

Region 3: Use either J-M Centurian or Coated Asbestos as the base felt, lapping each felt 2" over the preceding one and spot mop the full width under each felt with asphalt using a minimum of 10 lbs. per square.

Second: Starting at the low edge apply one 18" wide, then over that one 36" wide J-M Asbestos Finishing Felt. Following felts are to be applied full width overlapping the preceding felt by 19" in such manner that at least 2 plies of felt cover the base felt at any point. Broom each felt so that it shall be firmly and uniformly set without voids into hot Aquadam applied just before the felt at a minimum rate of 23 lbs. per square uniformly over the entire surface.

Third: Flood the surface with J-M Aquadam at a minimum rate of 60 lbs per square and while it is still hot embed therein an acceptable gravel at the rate of approximately 400 lbs per square or an acceptable slag at a rate of approximately 300 lbs per square.

Materials per 100 sq ft of roof area

CONCRETE PRIMER:	If required	1 gal
FELTS:	J-M Centurian or Coated Asbestos Base Felt	1 layer
	J-M Asbestos Finishing Felt	2 layers
BITUMEN:	J-M Aquadam	69 lbs*
SURFACING:	J-M Aquadam	60 lbs
	Gravel	400 lbs
	Slag	300 lbs

Approximately applied Weight Min: 471 lbs. Max: 597 lbs.

*Deduct 13 lbs. if Spot Mopped.

BUILT-UP ROOFS by Johns-Manville/Greenwood Plaza, Denver, Colorado 80217

For Regions 1, 2, & 3

Specification **No. 601-I** *for use over*

FESCO, FESCO-FOAM OR OTHER APPROVED INSULATION

*on inclines of up to **1/2"** per foot*

Johns-Manville
Gravel-Surface Asbestos BUILT-UP ROOFS

This specification is for use over Fesco, Fesco-Foam or any type of approved insulation which is not nailable and which offers suitable surface to receive the roof. This specification is not to be used over light-weight insulating concrete decks either poured or pre-cast, or over fill made of lightweight insulating concrete.

Preparation of Deck — For information on the Preparation of Deck and Roof Drainage, see section on ROOF DECKS, Specification Manual for Johns-Manville Built-Up Roofs.

Flashings

See section on FLASHINGS, Specification Manual for J-M Built-Up Roofs.

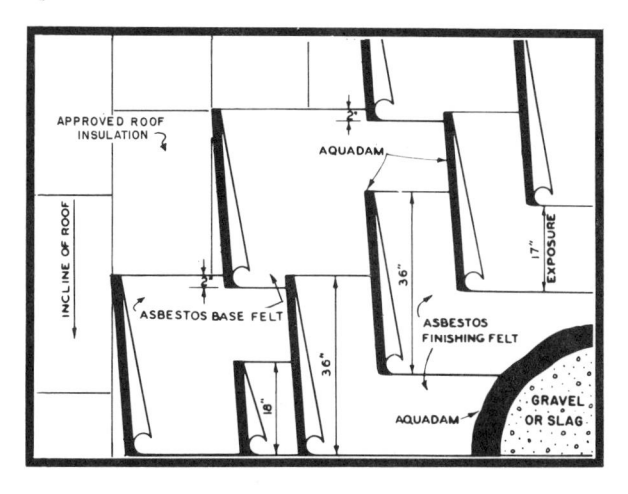

This specification is eligible for a 20-Year Guarantee only when, in the opinion of an authorized J-M Representative, all conditions listed in "General Instructions" of this Specification Manual have been met.

Application of Roofing

Regions 1 & 2: Use J-M Centurian as the base felt.

Region 3: Use either J-M Centurian or Coated Asbestos as the base felt.

First: Lap each felt 2" over the preceding one. Mop the full width under each felt with hot J-M asphalt using a minimum of 33 lbs. per square.

Second: Starting at the low edge apply one 18" wide, then over that one 36" wide J-M Asbestos Finishing Felt. Following felts are to be applied full width overlapping the preceding felt by 19" in such manner that at least 2 plies of felt cover the base felt at any point. Broom each felt so that it shall be firmly and uniformly set without voids into hot Aquadam applied just before the felt at a minimum rate of 23 lbs. per square uniformly over the entire surface.

Third: Flood the surface with J-M Aquadam at a minimum rate of 60 lbs per square and while it is still hot embed therein an acceptable gravel at the rate of approximately 400 lbs per square or an acceptable slag at a rate of approximately 300 lbs per square.

Materials per 100 sq ft of roof area

FELTS:	J-M Centurian or Coated Asbestos Base Felt	1 layer
	J-M Asbestos Finishing Felt	2 layers
BITUMEN:	J-M 190 Asphalt	33 lbs*
	J-M Aquadam	46 lbs
SURFACING:		
	J-M Aquadam	60 lbs
	Gravel	400 lbs
	Slag	300 lbs

Approximately applied Weight Min: 494 lbs. Max: 612 lbs.

*If over Fesco-Foam deduct 10 lbs.

BUILT-UP ROOFS *by Johns-Manville/Greenwood Plaza, Denver, Colorado 80217*

Regions 1, 2, & 3

*Specification **No. 630** for use over*

PLYWOOD OR OTHER NAILABLE DECKS

*on inclines of up to **3"** per foot*

Johns-Manville

Gravel-Surface Asbestos Felt BUILT-UP ROOFS

This specification is to be used over any type of structural deck (without insulation) which can receive and adequately retain nails or other types of mechanical fasteners as may be recommended by the deck manufacturer. Examples of such decks are wood, plywood, and some lightweight aggregate concrete decks* no lighter than 32 PCF density, or that provide proper nail retention.

*Ventsulation Felt must be the base felt over these decks.

Note: All information contained in "General Instructions" in the current Specification Manual for Johns-Manville Built-Up Roofs shall be considered part of this specification.

Flashings
See section on FLASHINGS, Specification Manual for J-M Built-Up Roofs.

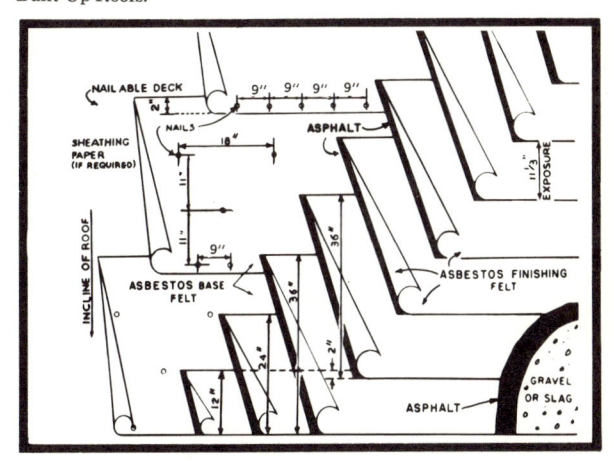

This specification is eligible for a 20-Year Guarantee only when, in the opinion of an authorized J-M Representative, all conditions listed in "General Instructions" of this Specification Manual have been met.

Application of Roofing

Over wood board decks one ply of sheathing paper must be used under the base felt next to the deck.

Regions 1 & 2: Use J-M Centurian as the base felt.

Region 3: Use either J-M Centurian or Coated Asbestos as the base felt.

First: Start at the low edge and working up the slope and perpendicular to the slope and lapping each felt 2" over the preceding one. Nail the laps at 9" centers and down the longitudinal center of each felt nail two rows of nails with rows spaced approximately 11" apart and nails staggered on approximately 18" centers. Use nails for fasteners appropriate to the type of deck.

Note: In Region 3 only if deck is plywood, base felt may be sprinkle mopped using 10 lbs. of asphalt per square.

Second: Starting at the low edge apply one 12" wide, then over that one 24" wide, then over both a full 36" wide Asbestos Finishing Felt, (Perforated). Following felts are to be applied full width overlapping the preceding felt by 24⅔" in such manner that at least 3 plies of felt cover the base felt at any point. Broom each felt so that it shall be firmly and uniformly set without voids into hot asphalt applied just before the felt at a minimum rate of 23 lbs per square uniformly over the entire surface.

On slopes 1" per foot or greater nail each felt at approximately 9" centers adjacent to the back edge.

Third: Flood the surface with the appropriate asphalt depending on the roof slope, at a minimum rate of 60 lbs per square and while it is still hot embed therein an acceptable gravel at the rate of approximately 400 lbs per square or an acceptable slag at a rate of approximately 300 lbs per square.

Materials per 100 sq ft of roof area

SHEATHING PAPER: Wood board decks only	1 layer
FELTS: J-M Centurian or Coated Asbestos Base Felt	1 layer
J-M Asbestos Finishing Felt (Perforated)	3 layers
BITUMEN: Aquadam—Slopes up to ½" per foot	69 lbs
190 Asphalt—Slopes ½" to 3" per foot	69 lbs
SURFACING: Appropriate Bitumen	60 lbs
Gravel	400 lbs
or	
Slag	300 lbs

Approximate Installed Weight Min: 499 lbs. Max: 617 lbs.

Nailing

All nails or other fasteners are to be driven through tin caps unless the nail or fastener has an integral flat cap no less than 1" across.

BUILT-UP ROOFS by Johns-Manville/Greenwood Plaza, Denver, Colorado 80217

Specification **No. 631** *for use over*

CONCRETE OR OTHER NON-NAILABLE DECKS

on inclines of up to 3" per foot

Johns-Manville
Gravel-Surface Asbestos Felt BUILT-UP ROOFS

This specification is for use over any type of structural deck which is not nailable and which offers suitable surface to receive the roof. Poured and pre-cast concrete decks require priming. This specification is not to be used over lightweight insulating concrete decks either poured or pre-cast, or over fill made of lightweight insulating concrete.

Note: All information contained in "General Instructions" in the current Specification Manual for Johns-Manville Built-Up Roofs shall be considered part of this specification.

Flashings

See section on FLASHINGS, Specification Manual for J-M Built-Up Roofs.

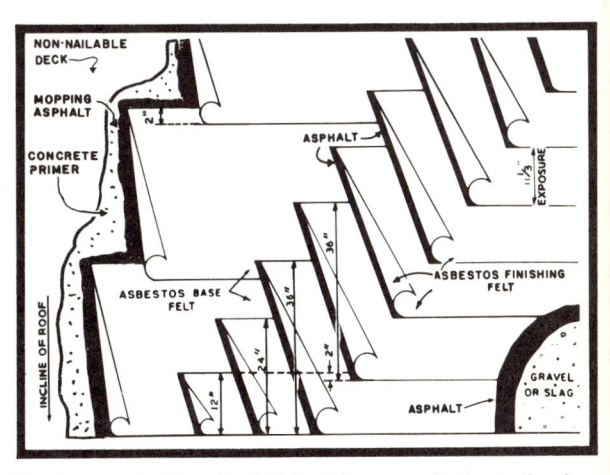

On slopes up to 1" apply finishing felts perpendicular to the slope starting at the low point of each slope. On slopes over 1" apply finishing felts parallel to the slope, nailing at the top of each run of felt on not over 9" centers. If run of felt exceeds 20' an additional line of nails shall be used at 20' intervals.

This specification is eligible for a 20-Year Guarantee only when, in the opinion of an authorized J-M Representative, all conditions listed in "General Instructions" of this Specification Manual have been met.

Application of Roofing

First: Regions 1 & 2: Use J-M Centurian as the base felt, lapping each felt 2″ over the preceding one and solidly mop the full width under each ply felt with asphalt using a minimum of 23 lbs. per square.

Region 3: Use either J-M Centurian or Coated Asbestos as the base felt, lapping each felt 2″ over the preceding one and spot mop the full width under each felt with asphalt using a minimum of 10 lbs. per square.

Second: Starting at the low edge apply one 12″ wide, then over that one 24″ wide, then over both a full 36″ wide Asbestos Finishing Felt (Perforated). Following felts are to be applied full width overlapping the preceding felt by 24⅔″ in such manner that at least 3 plies of felt cover the base felt at any point.

Broom each felt so that it shall be firmly and uniformly set without voids into hot Asphalt applied just before the felt at a minimum rate of 23 lbs per square uniformly over the entire surface.

On slopes over 1″ per foot, all felts shall be nailed at the top of each run of felt on not over 9″ centers. If run of felt exceeds 20′ an additional line of nails shall be used at 20′ intervals.

Third: Flood the surface with the appropriate Asphalt, depending on the roof slope, at a minimum rate of 60 lbs per square and while it is still hot embed therein an acceptable gravel at the rate of approximately 400 lbs per square or an acceptable slag at a rate of approximately 300 lbs per square.

Materials per 100 sq ft of roof area

CONCRETE PRIMER: If required	1 gal
FELTS: J-M Centurian or Coated Asbestos Base Felt	1 layer
J-M Asbestos Finishing Felt (Perforated)	3 layers
BITUMEN: Aquadam—Slopes up to ½″ per foot	92 lbs*
190 Asphalt—Slopes ½″ to 3″ per foot	92 lbs*
SURFACING: Appropriate Bitumen	60 lbs
Gravel	400 lbs
or	
Slag	300 lbs

Approximate Installed Weight Min: 522 lbs. Max: 658 lbs.

*Deduct 13 lbs. if Spot Mopped.

Nailing

Where nailing is required, nailing strips must be provided. All nails or other fasteners are to be driven through tin caps unless the nail or fastener has an integral flat cap no less than 1″ across.

BUILT-UP ROOFS by Johns-Manville/Greenwood Plaza, Denver, Colorado 80217

Regions 1, 2, & 3

Specification **No. 631-I** for use over

FESCO, FESCO-FOAM OR APPROVED INSULATION

on inclines of up to 3" per foot

Johns-Manville

Gravel-Surface Asbestos Felt BUILT-UP ROOFS

This specification is for use over Fesco, Fesco-Foam or any type of approved insulation which is not nailable and which offers suitable surface to receive the roof. This specification is not to be used over light-weight insulating concrete decks either poured or pre-cast, or over fill made of lightweight insulating concrete.

Note: All information contained in "General Instructions" in the current Specification Manual for Johns-Manville Built-Up Roofs shall be considered part of this specification.

Flashings

See section on FLASHINGS, Specification Manual for J-M Built-Up Roofs.

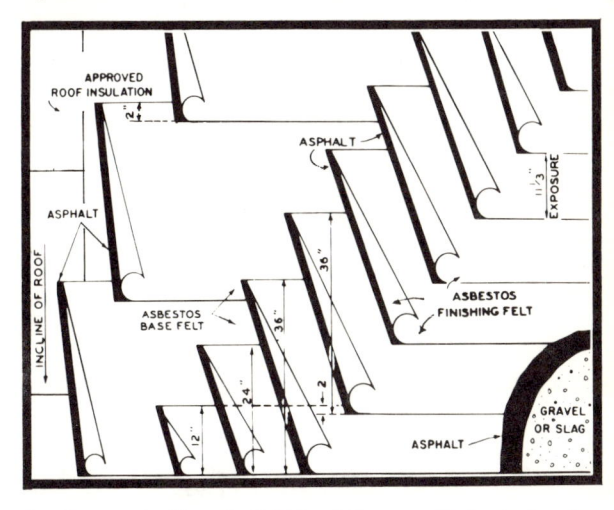

On slopes up to 1" apply finishing felts perpendicular to the slope starting at the low point of each slope. On slopes over 1" apply finishing felts parallel to the slope, nailing at the top of each run of felt on not over 9" centers. If run of felt exceeds 20' an additional line of nails shall be used at 20' intervals.

This specification is eligible for a 20-Year Guarantee only when, in the opinion of an authorized J-M Representative, all conditions listed in "General Instructions" of this Specification Manual have been met.

Application of Roofing

Regions 1 & 2: Use J-M Centurian as the base felt.

Region 3: Use either J-M Centurian or Coated Asbestos as the base felt.

First: Lap each felt 2" over the preceding one. Mop the full width under each felt with the appropriate hot J-M asphalt using a minimum of 33 lbs. per square.

Second: Starting at the low edge apply one 12" wide, then over that one 24" wide, then over both a full 36" wide Asbestos Finishing Felt (Perforated). Following felts are to be applied full width overlapping the preceding felt by 24⅔" in such manner that at least 3 plies of felt cover the base felt at any point.

Broom each felt so that it shall be firmly and uniformly set without voids into hot Asphalt applied just before the felt at a minimum rate of 23 lbs per square uniformly over the entire surface.

On slopes over 1" per foot, all felts shall be nailed at the top of each run of felt on not over 9" centers. If run of felt exceeds 20' an additional line of nails shall be used at 20' intervals.

Third: Flood the surface with the appropriate Asphalt, depending on the roof slope, at a minimum rate of 60 lbs per square and while it is still hot embed therein an acceptable gravel at the rate of approximately 400 lbs per square or an acceptable slag at a rate of approximately 300 lbs per square.

Materials per 100 sq ft of roof area

FELTS: J-M Centurian or Coated Asbestos Base Felt	1 layer
J-M Asbestos Finishing Felt (Perforated)	3 layers
BITUMEN: Aquadam—Slopes up to ½" per foot	69 lbs.
190 Asphalt—Slopes ½" to 3" per foot	102 lbs.*
SURFACING: Appropriate Bitumen	60 lbs
Gravel	400 lbs
or	
Slag	300 lbs

Approximate Installed Weight Min: 522 lbs. Max: 658 lbs.

*If over Fesco Foam deduct 10 lbs.

Nailing

Where nailing is required, nailing strips must be provided. All nails or other fasteners are to be driven through tin caps unless the nail or fastener has an integral flat cap no less than 1" across.

BUILT-UP ROOFS by Johns-Manville/Greenwood Plaza, Denver, Colorado 80217

Regions 1, 2, & 3

*Specification **No. 3000** for use over*
PLYWOOD OR OTHER NAILABLE DECKS
*on inclines of **1/2"** to **3"** per foot*

Johns-Manville
Gravel-Surface Asbestos BUILT-UP ROOFS

This specification is to be used over any type of structural deck (without insulation) which can receive and adequately retain nails or other types of mechanical fasteners as may be recommended by the deck manufacturer. Examples of such decks are wood, plywood, and some lightweight aggregate concrete decks* no lighter than 32 PCF density, or that provide proper nail retention.

*Ventsulation Felt must be the base felt over these decks.

Note: All information contained in "General Instructions" in the current Specification Manual for Johns-Manville Built-Up Roofs shall be considered part of this specification.

Flashings
See section on FLASHINGS, Specification Manual for J-M Built-Up Roofs.

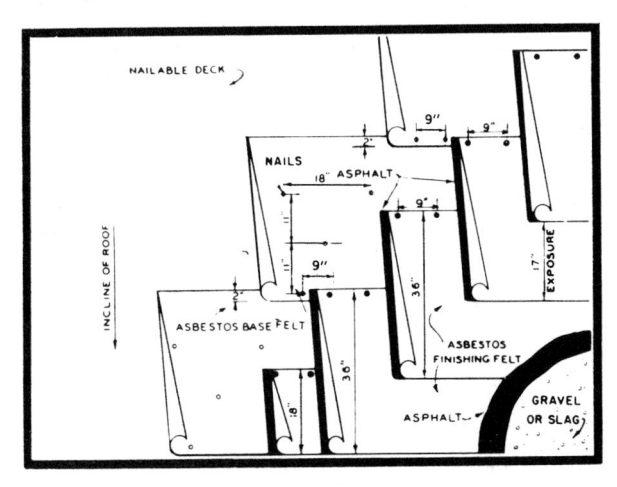

This specification is eligible for a 15-Year Guarantee in Region 1 and 20-Year Guarantee in Regions 2 & 3 only when, in the opinion of an authorized J-M Representative, all conditions listed in "General Instructions" of this Specification Manual have been met.

Application of Roofing

Over wood board decks one ply of sheathing paper must be used under the base felt next to the deck.

Regions 1 & 2: Use J-M Centurian as the base felt.

Region 3: Use either J-M Centurian or Coated Asbestos as the base felt.

First: Start at the low edge and working up the slope and perpendicular to the slope and lapping each felt 2" over the preceding one. Nail the laps at 9" centers and down the longitudinal center of each felt nail two rows of nails with rows spaced approximately 11" apart and nails staggered on approximately 18" centers. Use nails for fasteners appropriate to the type of deck.

Note: In Region 3 only if deck is plywood, base felt may be sprinkle mopped using 10 lbs. of asphalt per square.

Second: Starting at the low edge apply one 18" wide, then over that one 36" wide J-M Asbestos Finishing Felt. Following felts are to be applied full width overlapping the preceding felt by 19" in such manner that at least 2 plies of felt cover the base felt at any point. On slopes 1" per ft or greater nail each felt at approximately 9" centers adjacent to the back edge. Broom each felt so that it shall be firmly and uniformly set without voids into hot J-M 190 Asphalt applied just before the felt at a minimum rate of 23 lbs per square uniformly over the entire surface.

Third: Flood the surface with J-M 190 Asphalt at a minimum rate of 60 lbs per square and while it is still hot embed therein an acceptable gravel at the rate of approximately 400 lbs per square or an acceptable slag at a rate of approximately 300 lbs per square.

Materials per 100 sq ft of roof area

FELTS:	J-M Centurian or Coated Asbestos Base Felt	1 layer
	J-M Asbestos Finishing Felt	2 layers
BITUMEN:	J-M 190 Asphalt	46 lbs
	Sprinkle mopping to plywood	10 lbs
SURFACING:		
	J-M 190 Asphalt	60 lbs
	Gravel	400 lbs
	Slag	300 lbs

Approximately applied Weight Min: 471 lbs. Max: 589 lbs.

Nailing

All nails or other fasteners are to be driven through tin caps unless the nail or fastener has an integral flat cap no less than 1" across.

BUILT-UP ROOFS by Johns-Manville/Greenwood Plaza, Denver, Colorado 80217

r Regions 1, 2, & 3

Specification **No. 3001** *for use over*
CONCRETE OR OTHER NON-NAILABLE DECKS
on inclines of ***1/2"*** *to* ***3"*** *per foot*

Johns-Manville

Gravel-Surface Asbestos BUILT-UP ROOFS

This specification is for use over any type of structural deck which is not nailable and which offers suitable surface to receive the roof. Poured and pre-cast concrete decks require priming. This specification is not to be used over lightweight insulating concrete decks either poured or pre-cast, or over fill made of lightweight insulating concrete.

Note: All information contained in "General Instructions" in the current Specification Manual for Johns-Manville Built-Up Roofs shall be considered part of this specification.

Flashings

See section on FLASHINGS, Specification Manual for J-M Built-Up Roofs.

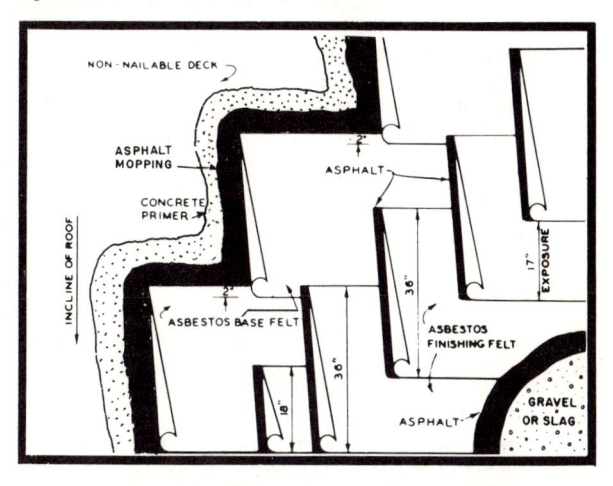

On slopes up to 1" apply finishing felts perpendicular to the slope starting at the low point of each slope. On slopes over 1" apply finishing felts parallel to the slope, nailing at the top of each run of felt on not over 9" centers. If run of felt exceeds 20' an additional line of nails shall be used at 20' intervals.

This specification is eligible for a 20-Year Guarantee only when, in the opinion of an authorized J-M Representative, all conditions listed in "General Instructions" of this Specification Manual have been met.

Application of Roofing

First: Regions 1 & 2: Use J-M Centurian as the base felt, lapping each felt 2" over the preceding one and solidly mop the full width under each ply felt with asphalt using a minimum of 23 lbs. per square.

Region 3: Use either J-M Centurian or Coated Asbestos as the base felt, lapping each felt 2" over the preceding one and spot mop the full width under each felt with asphalt using a minimum of 10 lbs. per square.

Second: Starting at the low edge (on slopes up to 1") apply one 18" wide, then over that one full 36" wide J-M Asbestos Finishing Felt. Following felts are to be applied full width overlapping the preceding felt by 19" in such manner that at least 2 plies of felt cover the base felt at any point. Broom each felt so that it shall be firmly and uniformly set without voids into hot J-M 190 Asphalt applied just before the felt at a minimum rate of 23 lbs. per square uniformly over the entire surface. On slopes over 1" per foot, all felts shall be nailed at the top of each run of felt on not over 9" centers. If run of felt exceeds 20' an additional line of nails shall be used at 20' intervals.

Third: Flood the surface with J-M 190 Asphalt at a minimum rate of 60 lbs per square and while it is still hot embed therein an acceptable gravel at the rate of approximately 400 lbs per square or an acceptable slag at a rate of approximately 300 lbs per square.

Materials per 100 sq ft of roof area

CONCRETE PRIMER:	If required	1 gal
FELTS:	J-M Centurian or Coated Asbestos Base Felt	1 layer
	J-M Asbestos Finishing Felt	2 layers
BITUMEN:	J-M 190 Asphalt	69 lbs.*
SURFACING:		
	J-M 190 Asphalt	60 lbs
	Gravel	400 lbs
	Slag	300 lbs

Approximately applied Weight Min: 471 lbs. Max: 597 lbs.

*Delete 13 lbs. if spot mopped.

Nailing

Where nailing is required, nailing strips must be provided. All nails or other fasteners are to be driven through tin caps unless the nail or fastener has an integral flat cap no less than 1" across.

BUILT-UP ROOFS by Johns-Manville/Greenwood Plaza, Denver, Colorado 80217

Regions 1, 2, & 3

Specification **No. 3001-I** *for use over*

FESCO, FESCO-FOAM OR OTHER APPROVED INSULATION

*on inclines of **1/2"** to **3"** per foot*

Johns-Manville

Gravel-Surface Asbestos BUILT-UP ROOFS

This specification is for use over Fesco, Fesco-Foam or any type of approved insulation which is not nailable and which offers suitable surface to receive the roof. This specification is not to be used over light-weight insulating concrete decks either poured or pre-cast, or over fill made of lightweight insulating concrete.

Note: All information contained in "General Instructions" in the current Specification Manual for Johns-Manville Built-Up Roofs shall be considered part of this specification.

Flashings

See section on FLASHINGS, Specification Manual for J-M Built-Up Roofs.

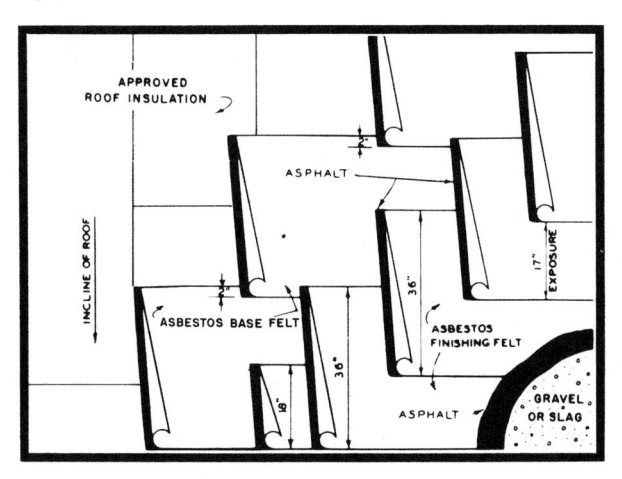

On slopes up to 1" apply finishing felts perpendicular to the slope starting at the low point of each slope. On slopes over 1" apply finishing felts parallel to the slope, nailing at the top of each run of felt on not over 9" centers. If run of felt exceeds 20' an additional line of nails shall be used at 20' intervals.

This specification is eligible for a 20-Year Guarantee only when, in the opinion of an authorized J-M Representative, all conditions listed in "General Instructions" of this Specification Manual have been met.

Application of Roofing

Regions 1 & 2: Use J-M Centurian as the base felt.

Region 3: Use either J-M Centurian or Coated Asbestos as the base felt.

First: Lap each felt 2" over the preceding one. Mop the full width under each felt with the appropriate hot J-M asphalt using a minimum of 33 lbs. per square.

Second: Starting at the low edge (on slopes up to 1") apply one 18" wide, then over that one full 36" wide J-M Asbestos Finishing Felt. Following felts are to be applied full width overlapping the preceding felt by 19" in such manner that at least 2 plies of felt cover the base felt at any point. Broom each felt so that it shall be firmly and uniformly set without voids into hot J-M 190 Asphalt applied just before the felt at a minimum rate of 23 lbs. per square uniformly over the entire surface. On slopes over 1" per foot, all felts shall be nailed at the top of each run of felt on not over 9" centers. If run of felt exceeds 20' an additional line of nails shall be used at 20' intervals.

Third: Flood the surface with J-M 190 Asphalt at a minimum rate of 60 lbs per square and while it is still hot embed therein an acceptable gravel at the rate of approximately 400 lbs per square or an acceptable slag at a rate of approximately 300 lbs per square.

Materials per 100 sq ft of roof area

FELTS:	J-M Centurian or Coated Asbestos Base Felt	1 layer
	J-M Asbestos Finishing Felt	2 layers
BITUMEN:	J-M 190 Asphalt	79 lbs *
SURFACING:		
	J-M 190 Asphalt	60 lbs
	Gravel	400 lbs
	Slag	300 lbs

Approximately applied Weight Min: 494 lbs. Max: 612 lbs.

*If over Fesco-Foam deduct 10 lbs.

Nailing

Where nailing is required, nailing strips must be provided. All nails or other fasteners are to be driven through tin caps unless the nail or fastener has an integral flat cap no less than 1" across.

BUILT-UP ROOFS by Johns-Manville/Greenwood Plaza, Denver, Colorado 80217

Regions 1, 2, & 3

Specification **No. 800** for use over

PLYWOOD OR OTHER NAILABLE DECKS

on inclines of up to 1/2" per foot

Johns-Manville

Gravel-Surface Organic Felt BUILT-UP ROOFS

This specification is to be used over any type of structural deck (without insulation) which can receive and adequately retain nails or other types of mechanical fasteners as may be recommended by the deck manufacturer. Examples of such decks are wood, plywood, and some lightweight aggregate concrete decks* no lighter than 32 PCF density, or that provide proper nail retention.

*Ventsulation Felt must be the base felt over these decks.

Note: All information contained in "General Instructions" in the current Specification Manual for Johns-Manville Built-Up Roofs shall be considered part of this specification.

Flashings
See section on FLASHINGS, Specification Manual for J-M Built-Up Roofs.

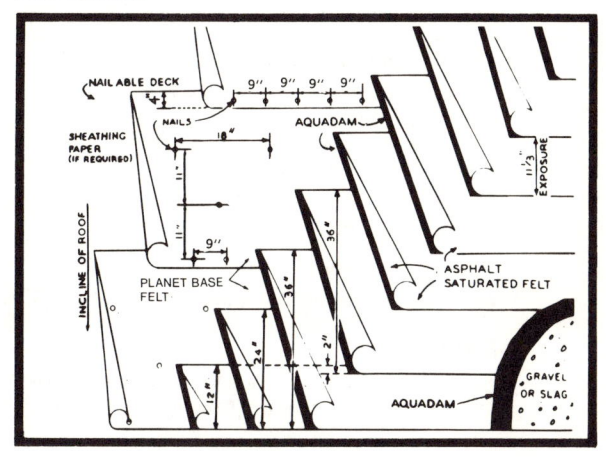

This specification is eligible for a 20-Year Guarantee only when, in the opinion of an authorized J-M Representative, all conditions listed in "General Instructions" of this Specification Manual have been met.

Application of Roofing

Over wood board decks one ply of sheathing paper must be used under the base felt next to the deck.

Regions 1, 2 & 3: Use J-M Planet as the base felt.

First: Start at the low edge and working up the slope and perpendicular to the slope and lapping each felt 4" over the preceding one. Nail the laps at 9" centers and down the longitudinal center of each felt nail two rows of nails with rows spaced approximately 11" apart and nails staggered on approximately 18" centers. Use nails for fasteners appropriate to the type of deck.

Note: In Region 3 only if deck is plywood, base felt may be sprinkle mopped using 10 lbs. of asphalt per square.

Second: Starting at the low edge apply one 12" wide, then over that one 24" wide, then over both a full 36" wide Asphalt Saturated Felt, (Perforated). Following felts are to be applied full width overlapping the preceding felt by 24⅔" in such manner that at least 3 plies of felt cover the base felt at any point. Broom each felt so that it shall be firmly and uniformly set without voids into hot Aquadam applied just before the felt at a minimum rate of 23 lbs per square uniformly over the entire surface.

Third: Flood the surface with the Aquadam at a minimum rate of 60 lbs per square and while it is still hot embed therein an acceptable gravel at the rate of approximately 400 lbs per square or an acceptable slag at a rate of approximately 300 lbs per square.

Materials per 100 sq ft of roof area

SHEATHING PAPER: Wood board decks only	1 layer
FELTS: J-M Planet Base Felt	1 layer
J-M Asphalt Saturated Felt (Perforated)	3 layers
BITUMEN: J-M Aquadam	69 lbs
SURFACING: J-M Aquadam	60 lbs
Gravel	400 lbs
or	
Slag	300 lbs

Approximate Installed Weight Min: 517 lbs. Max: 617 lbs.

Nailing

All nails or other fasteners are to be driven through tin caps unless the nail or fastener has an integral flat cap no less than 1" across.

BUILT-UP ROOFS *by Johns-Manville/Greenwood Plaza, Denver, Colorado 80217*

◄ Regions 1, 2, & 3

Specification **No. 801** *for use over*

CONCRETE OR OTHER NON-NAILABLE DECKS

on inclines of up to 1/2" per foot

Johns-Manville
Gravel-Surface Organic Felt BUILT-UP ROOFS

This specification is for use over any type of structural deck which is not nailable and which offers suitable surface to receive the roof. Poured and pre-cast concrete decks require priming. This specification is not to be used over lightweight insulating concrete decks either poured or pre-cast, or over fill made of lightweight insulating concrete.

Note: All information contained in "General Instructions" in the current Specification Manual for Johns-Manville Built-Up Roofs shall be considered part of this specification.

Flashings

See section on FLASHINGS, Specification Manual for J-M Built-Up Roofs.

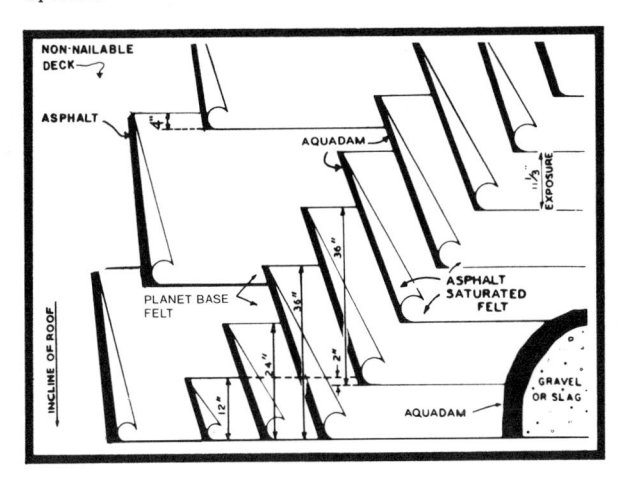

This specification is eligible for a 20-Year Guarantee only when, in the opinion of an authorized J-M Representative, all conditions listed in "General Instructions" of this Specification Manual have been met.

Application of Roofing

First: Regions 1 & 2: Use J-M Planet as the base felt, lapping each felt 4" over the preceding one and solidly mop the full width under each ply felt with Asphalt using a minimum of 23 lbs. per square.

Region 3: Use J-M Planet as the base felt, lapping each felt 4" over the preceding one and spot mop the full width under each felt with Asphalt using a minimum of 10 lbs. per square.

Second: Starting at the low edge apply one 12" wide, then over that one 24" wide, then over both a full 36" wide Asphalt Saturated Felt, (Perforated). Following felts are to be applied full width overlapping the preceding felt by 24⅔" in such manner that at least 3 plies of felt cover the base felt at any point. Broom each felt so that it shall be firmly and uniformly set without voids into hot Aquadam applied just before the felt at a minimum rate of 23 lbs per square uniformly over the entire surface.

Third: Flood the surface with the Aquadam at a minimum rate of 60 lbs per square and while it is still hot embed therein an acceptable gravel at the rate of approximately 400 lbs per square or an acceptable slag at a rate of approximately 300 lbs per square.

Materials per 100 sq ft of roof area

CONCRETE PRIMER: If required	1 gal
FELTS: J-M Planet Base Felt	1 layer
J-M Asphalt Saturated Felt (Perforated)	3 layers
BITUMEN: J-M 190 Asphalt	23 lbs*
J-M Aquadam	69 lbs
SURFACING: J-M Aquadam	60 lbs
Gravel	400 lbs
or	
Slag	300 lbs

Approximate Installed Weight Min: 522 lbs. Max: 658 lbs.
*Deduct 13 lbs. if spot mopped.

BUILT-UP ROOFS by Johns-Manville/Greenwood Plaza, Denver, Colorado 80217

Regions 1, 2, & 3

Specification **No. 801-I** *for use over*

FESCO, FESCO-FOAM OR APPROVED INSULATION

on inclines of up to 1/2" per foot

Johns-Manville

Gravel-Surface Organic Felt BUILT-UP ROOFS

This specification is for use over Fesco, Fesco-Foam or any type of approved insulation which is not nailable and which offers suitable surface to receive the roof. This specification is not to be used over light-weight insulating concrete decks either poured or pre-cast, or over fill made of lightweight insulating concrete.

Note: All information contained in "General Instructions" in the current Specification Manual for Johns-Manville Built-Up Roofs shall be considered part of this specification.

Flashings

See section on FLASHINGS, Specification Manual for J-M Built-Up Roofs.

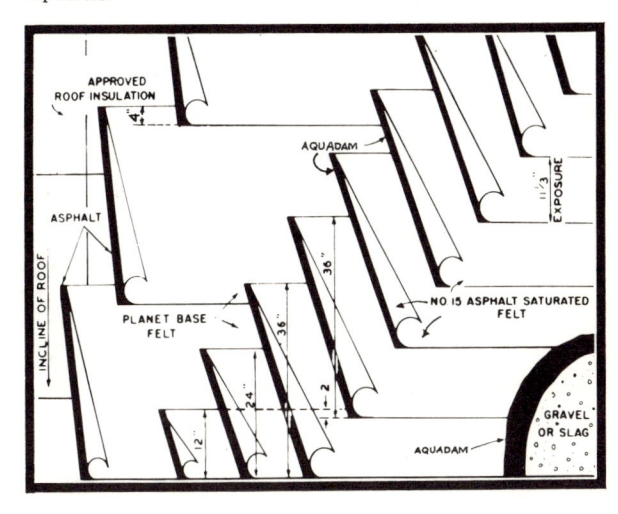

This specification is eligible for a 20-Year Guarantee only when, in the opinion of an authorized J-M Representative, all conditions listed in "General Instructions" of this Specification Manual have been met.

Application of Roofing

Regions 1, 2 & 3: Use J-M Planet as the base felt.

First: Lap each felt 4" over the preceding one. Mop the full width under each felt with the hot J-M Asphalt using a minimum of 33 lbs. per square.

Second: Starting at the low edge apply one 12" wide, then over that one 24" wide, then over both a full 36" wide Asphalt Saturated Felt, (Perforated). Following felts are to be applied full width overlapping the preceding felt by 24⅔" in such manner that at least 3 plies of felt cover the base felt at any point. Broom each felt so that it shall be firmly and uniformly set without voids into hot Aquadam applied just before the felt at a minimum rate of 23 lbs per square uniformly over the entire surface.

Third: Flood the surface with the Aquadam at a minimum rate of 60 lbs per square and while it is still hot embed therein an acceptable gravel at the rate of approximately 400 lbs per square or an acceptable slag at a rate of approximately 300 lbs per square.

Materials per 100 sq ft of roof area

FELTS:	J-M Planet Base Felt	1 layer
	J-M Asphalt Saturated Felt (Perforated)	3 layers
BITUMEN:	J-M Aquadam	69 lbs*
	J-M 190 Asphalt	33 lbs*
SURFACING:	Aquadam	60 lbs
	Gravel	400 lbs
	or	
	Slag	300 lbs

*if over Fesco-Foam deduct 10 lbs.

Approximate Installed Weight Min: 540 lbs. Max: 640 lbs.

Nailing

All nails or other fasteners are to be driven through tin caps unless the nail or fastener has an integral flat cap no less than 1" across.

BUILT-UP ROOFS by Johns-Manville/Greenwood Plaza, Denver, Colorado 80217

For Regions 1, 2, & 3

Specification No. 802 for use over

PLYWOOD OR OTHER NAILABLE DECKS

on inclines of up to 1/2" per foot

Johns-Manville
Gravel-Surface Organic Felt BUILT-UP ROOFS

This specification is to be used over any type of structural deck (without insulation) which can receive and adequately retain nails or other types of mechanical fasteners as may be recommended by the deck manufacturer. Examples of such decks are wood, plywood, and some lightweight aggregate concrete decks* no lighter than 32 PCF density, or that provide proper nail retention.

*Ventsulation Felt must be the base felt over these decks.

Note: All information contained in "General Instructions" in the current Specification Manual for Johns-Manville Built-Up Roofs shall be considered part of this specification.

Flashings
See section on FLASHINGS, Specification Manual for J-M Built-Up Roofs.

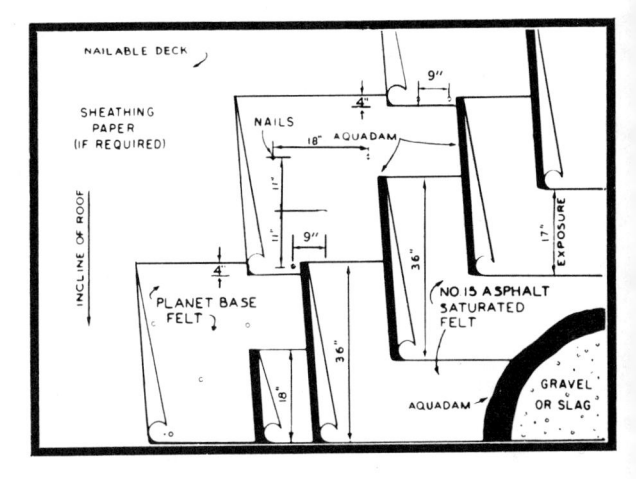

This specification is eligible for a 15-Year Guarantee only when, in the opinion of an authorized J-M Representative, all conditions listed in "General Instructions" of this Specification Manual have been met.

Application of Roofing

Over wood board decks one ply of sheathing paper must be used under the base felt next to the deck.

Regions 1, 2 & 3: Use J-M Planet as the base felt.

First: Start at the low edge and working up the slope and perpendicular to the slope and lapping each felt 4" over the preceding one. Nail the laps at 9" centers and down the longitudinal center of each felt nail two rows of nails with rows spaced approximately 11" apart and nails staggered on approximately 18" centers. Use nails for fasteners appropriate to the type of deck.

Note: In Region 3 only if deck is plywood, base felt may be sprinkle mopped using 10 lbs. of asphalt per square.

Second: Starting at the low edge apply one 18" wide, then over that one full 36" wide J-M No. 15 Asphalt Saturated Felt. (Perforated). Following felts are to be applied full width overlapping the preceding felt by 19" in such manner that at least 2 plies of felt cover the base felt at any point. Broom each felt so that it shall be firmly and uniformly set without voids into hot J-M Aquadam applied just before the felt at a minimum rate of 23 lbs per square uniformly over the entire surface.

Third: Flood the surface with J-M Aquadam at a minimum rate of 60 lbs per square and while it is still hot embed therein an acceptable gravel at the rate of approximately 400 lbs per square or an acceptable slag at a rate of approximately 300 lbs per square.

Materials per 100 sq ft of roof area

SHEATHING PAPER: Wood board decks only	1 layer
FELTS: J-M Planet Base Felt	1 layer
J-M Asphalt Saturated Felt (Perforated)	2 layers
BITUMEN: J-M Aquadam	46 lbs
Sprinkle mopping to plywood	10 lbs
SURFACING: J-M Aquadam	60 lbs
Gravel	400 lbs
or	
Slag	300 lbs

Approximate Installed Weight Min: 479 lbs. Max: 579 lbs.

Nailing

All nails or other fasteners are to be driven through tin caps unless the nail or fastener has an integral flat cap no less than 1" across.

BUILT-UP ROOFS by Johns-Manville/Greenwood Plaza, Denver, Colorado 80217

*Specification **No. 803** for use over*

CONCRETE OR OTHER NON-NAILABLE DECKS

on inclines of up to 1/2" per foot

Johns-Manville

Gravel-Surface Organic Felt BUILT-UP ROOFS

This specification is for use over any type of structural deck which is not nailable and which offers suitable surface to receive the roof. Poured and pre-cast concrete decks require priming. This specification is not to be used over lightweight insulating concrete decks either poured or pre-cast, or over fill made of lightweight insulating concrete.

Note: All information contained in "General Instructions" in the current Specification Manual for Johns-Manville Built-Up Roofs shall be considered part of this specification.

Flashings
See section on FLASHINGS, Specification Manual for J-M Built-Up Roofs.

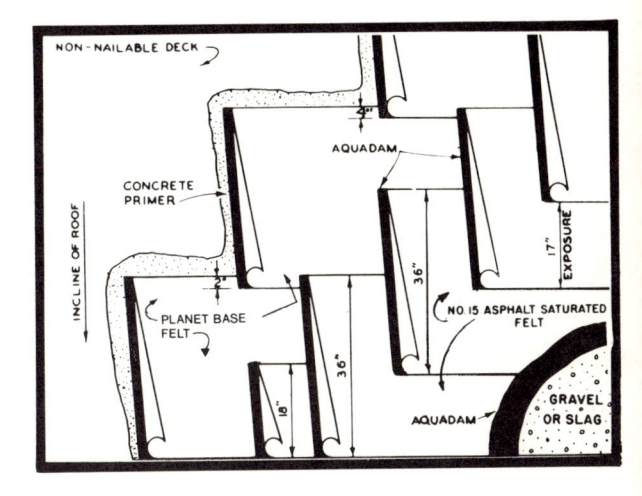

This specification is eligible for a 15-Year Guarantee only when, in the opinion of an authorized J-M Representative, all conditions listed in "General Instructions" of this Specification Manual have been met.

Application of Roofing

First: Regions 1 & 2: Use J-M Planet as the base felt, lapping each felt 4" over the preceding one and solidly mop the full width under each ply felt with asphalt using a minimum of 23 lbs. per square.

Region 3: Use J-M Planet as the base felt, lapping each felt 4" over the preceding one and spot mop the full width under each felt with asphalt using a minimum of 10 lbs. per square.

Second: Starting at the low edge apply one 18" wide, then over that one full 36" wide J-M No. 15 Asphalt Saturated Felt. (Perforated). Following felts are to be applied full width overlapping the preceding felt by 19" in such manner that at least 2 plies of felt cover the base felt at any point. Broom each felt so that it shall be firmly and uniformly set without voids into hot J-M Aquadam applied just before the felt at a minimum rate of 23 lbs per square uniformly over the entire surface.

Third: Flood the surface with J-M Aquadam at a minimum rate of 60 lbs per square and while it is still hot embed therein an acceptable gravel at the rate of approximately 400 lbs per square or an acceptable slag at a rate of approximately 300 lbs per square.

Materials per 100 sq ft of roof area

CONCRETE PRIMER: If required	1 gal
FELTS: J-M Planet Base Felt	1 layer
J-M Asphalt Saturated Felt (Perforated)	2 layers
BITUMEN: J-M 190 Asphalt	23 lbs.*
J-M Aquadam	46 lbs
SURFACING: J-M Aquadam	60 lbs
Gravel	400 lbs
or	
Slag	300 lbs

*Deduct 13 lbs. if spot mopped.

Approximate Installed Weight Min: 502 lbs. Max: 602 lbs.

BUILT-UP ROOFS by Johns-Manville/Greenwood Plaza, Denver, Colorado 80217

or Regions 1, 2, & 3

Specification No. 803-I for use over

FESCO, FESCO-FOAM OR OTHER APPROVED INSULATION

on inclines of up to 1/2" per foot

Johns-Manville

Gravel-Surface Organic Felt BUILT-UP ROOFS

This specification is for use over Fesco, Fesco-Foam or any type of approved insulation which is not nailable and which offers suitable surface to receive the roof. This specification is not to be used over light-weight aggregate concrete decks either poured or pre-cast, or over fill made of lightweight insulation concrete.

Preparation of Deck — For information on the Preparation of Deck and Roof Drainage, see section on ROOF DECKS, Specification Manual for Johns-Manville Built-Up Roofs.

Flashings

See section on FLASHINGS, Specification Manual for J-M Built-Up Roofs.

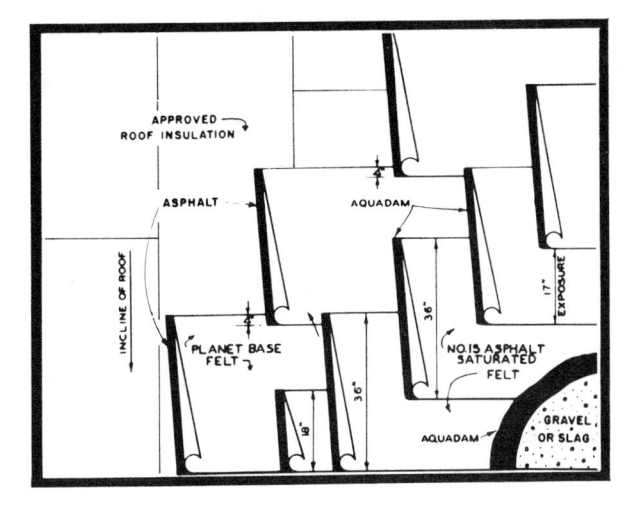

This specification is eligible for a 15-Year Guarantee only when, in the opinion of an authorized J-M Representative, all conditions listed in "General Instructions" of this Specification Manual have been met.

Application of Roofing

Regions 1, 2 & 3: Use J-M Planet as the base felt.

First: Lap each felt 4" over the preceding one. Mop the full width under each felt with the appropriate hot J-M asphalt using a minimum of 33 lbs. per square.

Second: Starting at the low edge apply one 18" wide, then over that one full 36" wide J-M No. 15 Asphalt Saturated Felt. (Perforated). Following felts are to be applied full width overlapping the preceding felt by 19" in such manner that at least 2 plies of felt cover the base felt at any point. Broom each felt so that it shall be firmly and uniformly set without voids into hot J-M Aquadam applied just before the felt at a minimum rate of 23 lbs per square uniformly over the entire surface.

Third: Flood the surface with J-M Aquadam at a minimum rate of 60 lbs per square and while it is still hot embed therein an acceptable gravel at the rate of approximately 400 lbs per square or an acceptable slag at a rate of approximately 300 lbs per square.

Materials per 100 sq ft of roof area

CONCRETE PRIMER:	If required	1 gal
FELTS:	J-M Planet Base Felt	1 layer
	J-M Asphalt Saturated Felt (Perforated)	2 layers
BITUMEN:	J-M 190 Asphalt	33 lbs.*
	J-M Aquadam	46 lbs
SURFACING:	J-M Aquadam	60 lbs
	Gravel	400 lbs
	or	
	Slag	300 lbs

*Deduct 10 lbs. over Fesco-Foam.

Approximate Installed Weight Min: 502 lbs. Max: 602 lbs.

BUILT-UP ROOFS by Johns-Manville/Greenwood Plaza, Denver, Colorado 80217

Regions 1, 2, & 3

Specification **No. 900** *for use over*

PLYWOOD OR OTHER NAILABLE DECKS

on inclines of 1/2" to 3" per foot

Johns-Manville
Gravel-Surface Organic Felt BUILT-UP ROOFS

This specification is to be used over any type of structural deck (without insulation) which can receive and adequately retain nails or other types of mechanical fasteners as may be recommended by the deck manufacturer. Examples of such decks are wood, plywood, and some lightweight aggregate concrete decks* no lighter than 32 PCF density, or that provide proper nail retention.

*Ventsulation Felt must be the base felt over these decks.

Note: All information contained in "General Instructions" in the current Specification Manual for Johns-Manville Built-Up Roofs shall be considered part of this specification.

Flashings
See section on FLASHINGS, Specification Manual for J-M Built-Up Roofs.

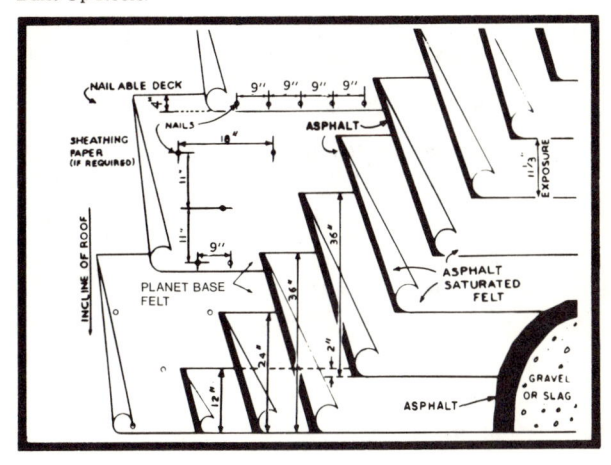

This specification is eligible for a 20-Year Guarantee only when, in the opinion of an authorized J-M Representative, all conditions listed in "General Instructions" of this Specification Manual have been met.

Application of Roofing

Over wood board decks one ply of sheathing paper must be used under the base felt next to the deck.

Regions 1, 2 & 3: Use J-M Planet as the base felt.

First: Start at the low edge and working up the slope and perpendicular to the slope and lapping each felt 4″ over the preceding one. Nail the laps at 9″ centers and down the longitudinal center of each felt nail two rows of nails with rows spaced approximately 11″ apart and nails staggered on approximately 18″ centers. Use nails for fasteners appropriate to the type of deck.

Note: In Region 3 only if deck is plywood, base felt may be sprinkle mopped using 10 lbs. of asphalt per square.

Second: Starting at the low edge apply one 12″ wide, then over that one 24″ wide, then over both a full 36″ wide Asphalt Saturated Felt, (Perforated). Following felts are to be applied full width overlapping the preceding felt by 24⅔″ in such manner that at least 3 plies of felt cover the base felt at any point. Broom each felt so that it shall be firmly and uniformly set without voids into hot Asphalt applied just before the felt at a minimum rate of 23 lbs per square uniformly over the entire surface.

On slopes 1″ per foot or greater nail each felt at approximately 9″ centers adjacent to the back edge.

Third: Flood the surface with the asphalt at a minimum rate of 60 lbs per square and while it is still hot embed therein an acceptable gravel at the rate of approximately 400 lbs per square or an acceptable slag at a rate of approximately 300 lbs per square.

Materials per 100 sq ft of roof area

SHEATHING PAPER: Wood board decks only	1 layer
FELTS: J-M Planet Base Felt	1 layer
J-M Asphalt Saturated Felt (Perforated)	3 layers
BITUMEN: J-M 190 Asphalt	69 lbs
SURFACING: J-M 190 Asphalt	60 lbs
Gravel	400 lbs
or	
Slag	300 lbs

Approximate Installed Weight Min: 517 lbs. Max: 617 lbs.

Nailing

All nails or other fasteners are to be driven through tin caps unless the nail or fastener has an integral flat cap no less than 1″ across.

BUILT-UP ROOFS by Johns-Manville/Greenwood Plaza, Denver, Colorado 80217

for Regions 1, 2, & 3

*Specification **No. 901** for use over*

CONCRETE OR OTHER NON-NAILABLE DECKS

*on inclines of **1/2"** to **3"** per foot*

Johns-Manville
Gravel-Surface Organic Felt BUILT-UP ROOFS

This specification is for use over any type of structural deck which is not nailable and which offers suitable surface to receive the roof. Poured and pre-cast concrete decks require priming. This specification is not to be used over lightweight insulating concrete decks either poured or pre-cast, or over fill made of lightweight insulating concrete.

Note: All information contained in "General Instructions" in the current Specification Manual for Johns-Manville Built-Up Roofs shall be considered part of this specification.

Flashings

See section on FLASHINGS, Specification Manual for J-M Built-Up Roofs.

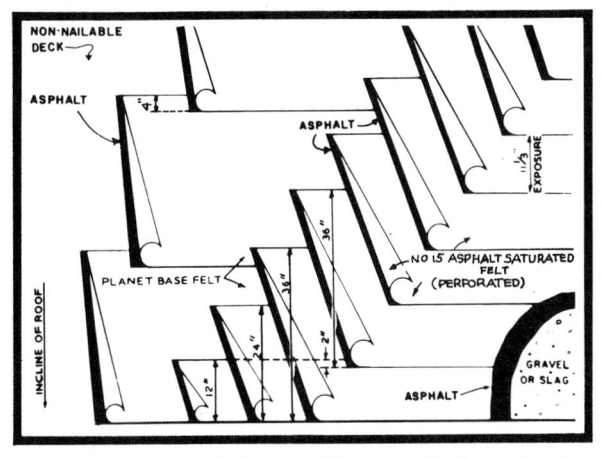

On slopes up to 1" apply finishing felts perpendicular to the slope starting at the low point of each slope. On slopes over 1" apply finishing felts parallel to the slope, nailing at the top of each run of felt on not over 9" centers. If run of felt exceeds 20' an additional line of nails shall be used at 20' intervals.

This specification is eligible for a 20-Year Guarantee only when, in the opinion of an authorized J-M Representative, all conditions listed in "General Instructions" of this Specification Manual have been met.

Application of Roofing

First: Regions 1 & 2: Use J-M Planet as the base felt, lapping each felt 4" over the preceding one and solidly mop the full width under each ply felt with asphalt using a minimum of 23 lbs. per square.

Region 3: Use J-M Planet as the base felt, lapping each felt 4" over the preceding one and spot mop the full width under each felt with asphalt using a minimum of 10 lbs. per square.

Second: Starting at the low edge apply one 12" wide, then over that one 24" wide, then over both a full 36" wide Asphalt Saturated Felt, (Perforated). Following felts are to be applied full width overlapping the preceding felt by 24⅔" in such manner that at least 3 plies of felt cover the base felt at any point. Broom each felt so that it shall be firmly and uniformly set without voids into hot Asphalt applied just before the felt at a minimum rate of 23 lbs per square uniformly over the entire surface.

On slopes over 1" per foot, all felts shall be nailed at the top of each run of felt on not over 9" centers. If run of felt exceeds 20' an additional line of nails shall be used at 20' intervals.

Third: Flood the surface with the Asphalt at a minimum rate of 60 lbs per square and while it is still hot embed therein an acceptable gravel at the rate of approximately 400 lbs per square or an acceptable slag at a rate of approximately 300 lbs per square.

Materials per 100 sq ft of roof area

CONCRETE PRIMER: If required	1 gal
FELTS: J-M Planet Base Felt	1 layer
J-M Asphalt Saturated Felt (Perforated)	3 layers
BITUMEN: J-M 190 Asphalt	92 lbs*
SURFACING: J-M 190 Asphalt	60 lbs
Gravel	400 lbs
or	
Slag	300 lbs

Approximate Installed Weight Min: 540 lbs. Max: 640 lbs.

*Deduct 13 lbs. if Spot Mopped.

Nailing

Where nailing is required, nailing strips must be provided. All nails or other fasteners are to be driven through tin caps unless the nail or fastener has an integral flat cap no less than 1" across.

BUILT-UP ROOFS by Johns-Manville/Greenwood Plaza, Denver, Colorado 80217

or Regions 1, 2, & 3

Specification **No. 901-I** *for use over*

FESCO, FESCO-FOAM OR APPROVED INSULATION

*on inclines of **1/2"** to **3"** per foot*

Johns-Manville

Gravel-Surface Organic Felt BUILT-UP ROOFS

This specification is for use over Fesco, Fesco-Foam or any type of approved insulation which is not nailable and which offers suitable surface to receive the roof. This specification is not to be used over light-weight insulating concrete decks either poured or pre-cast, or over fill made of lightweight insulating concrete.

Note: All information contained in "General Instructions" in the current Specification Manual for Johns-Manville Built-Up Roofs shall be considered part of this specification.

Flashings

See section on FLASHINGS, Specification Manual for J-M Built-Up Roofs.

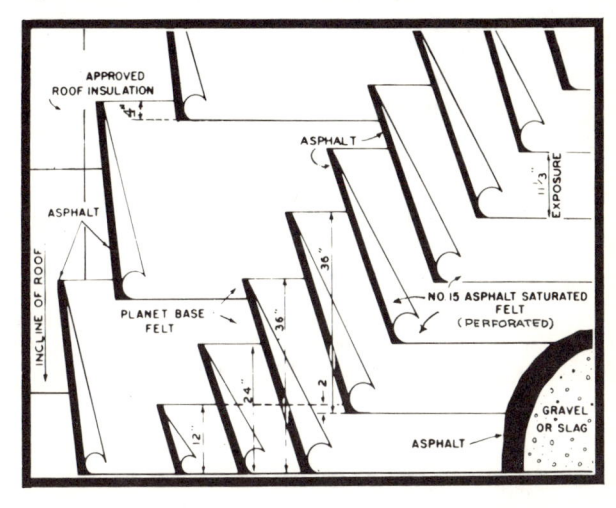

On slopes up to 1" apply finishing felts perpendicular to the slope starting at the low point of each slope. On slopes over 1" apply finishing felts parallel to the slope, nailing at the top of each run of felt on not over 9" centers. If run of felt exceeds 20' an additional line of nails shall be used at 20' intervals.

*This specification is eligible for a 20-Year Guarantee only when, in the opinion of an authorized J-M Representative, all conditions listed in "General Instructions" of this Specification **Manual have been** met.*

149

Application of Roofing

Regions 1, 2 & 3: Use J-M Planet as the base felt.

First: Lap each felt 4" over the preceding one. Mop the full width under each felt with the hot J-M Asphalt using a minimum of 33 lbs. per square.

Second: Starting at the low edge apply one 12" wide, then over that one 24" wide, then over both a full 36" wide Asphalt Saturated Felt, (Perforated). Following felts are to be applied full width overlapping the preceding felt by 24⅔" in such manner that at least 3 plies of felt cover the base felt at any point. Broom each felt so that it shall be firmly and uniformly set without voids into hot Asphalt applied just before the felt at a minimum rate of 23 lbs per square uniformly over the entire surface.

On slopes over 1" per foot, all felts shall be nailed at the top of each run of felt on not over 9" centers. If run of felt exceeds 20' an additional line of nails shall be used at 20' intervals.

Third: Flood the surface with the Asphalt at a minimum rate of 60 lbs per square and while it is still hot embed therein an acceptable gravel at the rate of approximately 400 lbs per square or an acceptable slag at a rate of approximately 300 lbs per square.

Materials per 100 sq ft of roof area

FELTS: J-M Planet Base Felt	1 layer
J-M Asphalt Saturated Felt (Perforated)	3 layers
BITUMEN: J-M 190 Asphalt	102 lbs. *
SURFACING: J-M 190 Asphalt	60 lbs
Gravel	400 lbs
or	
Slag	300 lbs

Approximate Installed Weight Min: 540 lbs. Max: 640 lbs.

*Deduct 10 lbs. if over Fesco-Foam.

Nailing

Where nailing is required, nailing strips must be provided. All nails or other fasteners are to be driven through tin caps unless the nail or fastener has an integral flat cap no less than 1" across.

BUILT-UP ROOFS by Johns-Manville/Greenwood Plaza, Denver, Colorado 80217

Specification **No. 902** for use over

PLYWOOD OR OTHER NAILABLE DECKS

on inclines of **1/2"** to **3"** per foot

Johns-Manville
Gravel-Surface Organic Felt BUILT-UP ROOFS

This specification is to be used over any type of structural deck (without insulation) which can receive and adequately retain nails or other types of mechanical fasteners as may be recommended by the deck manufacturer. Examples of such decks are wood, plywood, and some lightweight aggregate concrete decks* no lighter than 32 PCF density, or that provide proper nail retention.

*Ventsulation Felt must be the base felt over these decks.

Note: All information contained in "General Instructions" in the current Specification Manual for Johns-Manville Built-Up Roofs shall be considered part of this specification.

Flashings
See section on FLASHINGS, Specification Manual for J-M Built-Up Roofs.

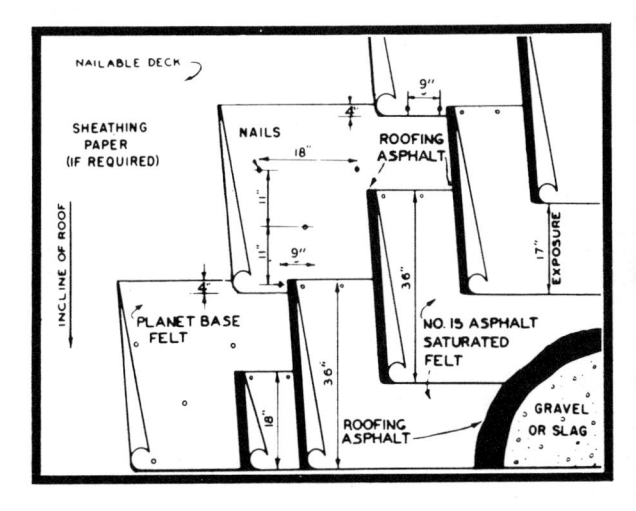

This specification is eligible for a 15-Year Guarantee only when, in the opinion of an authorized J-M Representative, all conditions listed in "General Instructions" of this Specification Manual have been met.

Application of Roofing

Over wood board decks one ply of sheathing paper must be used under the base felt next to the deck.

Regions 1, 2 & 3: Use J-M Planet as the base felt.

First: Start at the low edge and working up the slope and perpendicular to the slope and lapping each felt 4" over the preceding one. Nail the laps at 9" centers and down the longitudinal center of each felt nail two rows of nails with rows spaced approximately 11" apart and nails staggered on approximately 18" centers. Use nails for fasteners appropriate to the type of deck.

Note: In Region 3 only if deck is plywood, base felt may be sprinkle mopped using 10 lbs. of asphalt per square.

Second: Starting at the low edge apply one 18" wide, then over that one full 36" wide J-M No. 15 Asphalt Saturated Felt (Perforated). Following felts are to be applied full width overlapping the preceding felt by 19" in such manner that at least 2 plies of felt cover the base felt at any point. On slopes 1" per foot or greater nail each felt at approximately 9" centers adjacent to the back edge. Broom each felt so that it shall be firmly and uniformly set without voids into hot J-M 190 Asphalt applied just before the felt at a minimum rate of 23 lbs per square uniformly over the entire surface.

Third: Flood the surface with J-M 190 Asphalt at a minimum rate of 60 lbs per square and while it is still hot embed therein an acceptable gravel at the rate of approximately 400 lbs per square or an acceptable slag at a rate of approximately 300 lbs per square.

Materials per 100 sq ft of roof area

SHEATHING PAPER: Wood board decks only	1 layer
FELTS: J-M Planet Base Felt	1 layer
J-M Asphalt Saturated Felt (Perforated)	2 layers
BITUMEN: J-M 190 Asphalt	46 lbs
Sprinkle mopping to plywood	10 lbs
SURFACING: J-M 190 Asphalt	60 lbs
Gravel	400 lbs
or	
Slag	300 lbs

Approximate Installed Weight Min: 479 lbs. Max: 579 lbs.

Nailing

All nails or other fasteners are to be driven through tin caps unless the nail or fastener has an integral flat cap no less than 1" across.

BUILT-UP ROOFS *by Johns-Manville/Greenwood Plaza, Denver, Colorado 80217*

*Specification **No. 903** for use over*

CONCRETE OR OTHER NON-NAILABLE DECKS

on inclines of 1/2" to 3" per foot

Johns-Manville
Gravel-Surface Organic Felt BUILT-UP ROOFS

This specification is for use over any type of structural deck which is not nailable and which offers suitable surface to receive the roof. Poured and pre-cast concrete decks require priming. This specification is not to be used over lightweight insulating concrete decks either poured or pre-cast, or over fill made of lightweight insulating concrete.

Note: All information contained in "General Instructions" in the current Specification Manual for Johns-Manville Built-Up Roofs shall be considered part of this specification.

Flashings
See section on FLASHINGS, Specification Manual for J-M Built-Up Roofs.

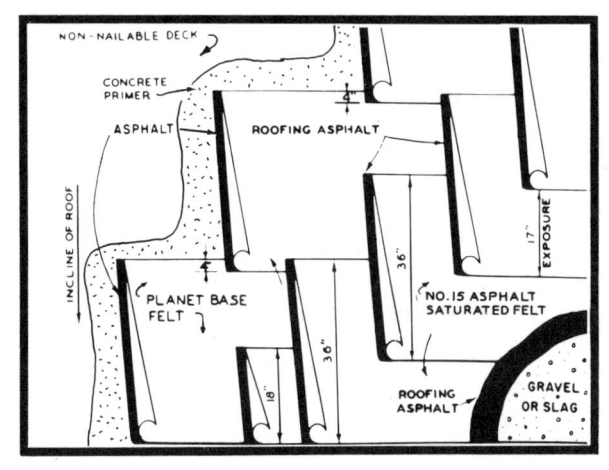

On slopes up to 1" apply felts perpendicular to the slope starting at the low point of each slope. On slopes over 1" apply felts parallel to the slope, nailing at the top of each run of felt on not over 9" centers. If run of felt exceeds 20' an additional line of nails shall be used at 20' intervals.

This specification is eligible for a 15-Year Guarantee only when, in the opinion of an authorized J-M Representative, all conditions listed in "General Instructions" of this Specification Manual have been met.

Application of Roofing

First: Regions 1 & 2: Use J-M Planet as the base felt, lapping each felt 4" over the preceding one and solidly mop the full width under each ply felt with asphalt using a minimum of 23 lbs. per square.

Region 3: Use J-M Planet as the base felt, lapping each felt 4" over the preceding one and spot mop the full width under each felt with asphalt using a minimum of 10 lbs. per square.

Second: Starting at the low edge (on slopes up to 1") apply one 18" wide, then over that one full 36" wide J-M No. 15 Asphalt Saturated Felt (Perforated). Following felts are to be applied full width overlapping the preceding felt by 19" in such manner that at least 2 plies of felt cover the base felt at any point. Broom each felt so that it shall be firmly and uniformly set without voids into hot J-M 190 Asphalt applied just before the felt at a minimum rate of 23 lbs per square uniformly over the entire surface. On slopes over 1" per foot all felts shall be nailed at the top of each run of felt on not over 9" centers. If run of felt exceeds 20' an additional line of nails shall be used at 20' intervals.

Third: Flood the surface with J-M 190 Asphalt at a minimum rate of 60 lbs per square and while it is still hot embed therein an acceptable gravel at the rate of approximately 400 lbs per square or an acceptable slag at a rate of approximately 300 lbs per square.

Materials per 100 sq ft of roof area

CONCRETE PRIMER: If required	1 gal
FELTS: J-M Planet Base Felt	1 layer
J-M Asphalt Saturated Felt (Perforated)	2 layers
BITUMEN: J-M 190 Asphalt	69 lbs*
SURFACING: J-M 190 Asphalt	60 lbs
Gravel	400 lbs
or	
Slag	300 lbs

*Deduct 13 lbs. if spot mopped.

Approximate Installed Weight Min: 502 lbs. Max: 602 lbs.

Nailing

Where nailing is required, nailing strips must be provided. All nails or other fasteners are to be driven through tin caps unless the nail or fastener has an integral flat cap no less than 1" across.

BUILT-UP ROOFS by Johns-Manville/Greenwood Plaza, Denver Colorado 80217

Regions 1, 2, & 3

Specification **No. 903 -I** *for use over*

FESCO, FESCO-FOAM OR OTHER APPROVED INSULATION

on inclines of 1/2" to 3" per foot

Johns-Manville
Gravel-Surface Organic Felt BUILT-UP ROOFS

This specification is for use over Fesco, Fesco-Foam or any type of approved insulation which is not nailable and which offers suitable surface to receive the roof. This specification is not to be used over light-weight insulating concrete decks either poured or pre-cast, or over fill made of lightweight insulating concrete.

Preparation of Deck — For information on the Preparation of Deck and Roof Drainage, see section on ROOF DECKS, Specification Manual for Johns-Manville Built-Up Roofs.

Flashings

See section on FLASHINGS, Specification Manual for J-M Built-Up Roofs.

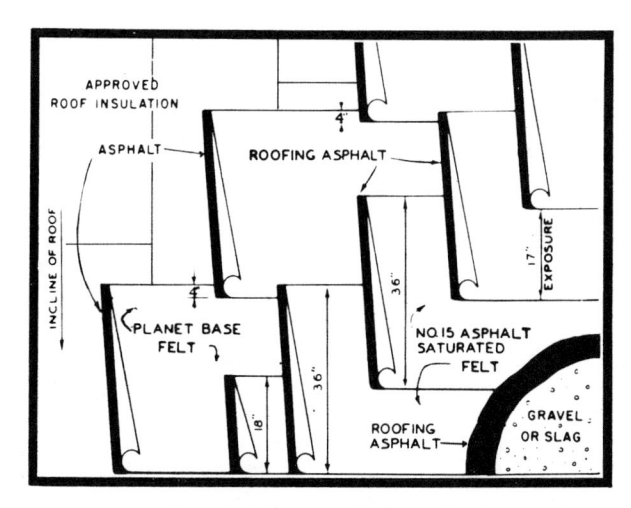

On slopes up to 1" apply felts perpendicular to the slope starting at the low point of each slope. On slopes over 1" apply felts parallel to the slope, nailing at the top of each run of felt on not over 9" centers. If run of felt exceeds 20' an additional line of nails shall be used at 20' intervals.

This specification is eligible for a 15-Year Guarantee only when, in the opinion of an authorized J-M Representative, all conditions listed in "General Instructions" of this Specification Manual have been met.

Application of Roofing

Regions 1, 2 & 3: Use J-M Planet as the base felt.

First: Lap each felt 4" over the preceding one. Mop the full width under each felt with the appropriate hot J-M asphalt using a minimum of 33 lbs. per square.

Second: Starting at the low edge (on slopes up to 1") apply one 18" wide, then over that one full 36" wide J-M No. 15 Asphalt Saturated Felt (Perforated). Following felts are to be applied full width overlapping the preceding felt by 19" in such manner that at least 2 plies of felt cover the base felt at any point. Broom each felt so that it shall be firmly and uniformly set without voids into hot J-M 190 Asphalt applied just before the felt at a minimum rate of 23 lbs per square uniformly over the entire surface. On slopes over 1" per foot all felts shall be nailed at the top of each run of felt on not over 9" centers. If run of felt exceeds 20' an additional line of nails shall be used at 20' intervals.

Third: Flood the surface with J-M 190 Asphalt at a minimum rate of 60 lbs per square and while it is still hot embed therein an acceptable gravel at the rate of approximately 400 lbs per square or an acceptable slag at a rate of approximately 300 lbs per square.

Materials per 100 sq ft of roof area

CONCRETE PRIMER: If required	1 gal
FELTS: J-M Planet Base Felt	1 layer
J-M Asphalt Saturated Felt (Perforated)	2 layers
BITUMEN: J-M 190 Asphalt	79 lbs.*
SURFACING: J-M 190 Asphalt	60 lbs
Gravel	400 lbs
or	
Slag	300 lbs

*Deduct 10 lbs. if over Fesco-Foam.

Approximate Installed Weight Min: 502 lbs. Max: 602 lbs.

Nailing

Where nailing is required, nailing strips must be provided. All nails or other fasteners are to be driven through tin caps unless the nail or fastener has an integral flat cap no less than 1" across.

BUILT-UP ROOFS by Johns-Manville/Greenwood Plaza, Denver, Colorado 80217

Johns-Manville
Ventsulation ASBESTOS FELT

*Provides
"a built-up roof that breathes"*

Johns-Manville
Ventsulation ASBESTOS FELT

Ventsulation Felt is designed so that air and moisture can "ventilate out" of the roof assembly both during construction and throughout the life of the roof. This makes it possible to eliminate the blistering, cracking, and premature failure once caused by sealed-in air and moisture that had no way to escape. In other words, Ventsulation Felt can provide . . . a Built-Up Roof That "breathes".

Ventsulation Felt is a heavy asbestos felt that is asphalt-saturated and coated. Mineral granules are embedded in the underside and then the underside is pressed into a "waffle" shape. The felt is applied to the deck waffle side down thus providing hundreds of small channels between the felt and the deck for free outward passage of air and moisture.

Ventsulation Felt can be applied over any type of roof deck that is firm, dry, clean and properly graded to outlets. Then the balance of the roofing membrane and flashings are installed according to their respective specifications.

Ventsulation Felt will take the place of the base felt in all specifications requiring a base felt.

*Specification **No. VS-1** for use over*
UNINSULATED DECKS

Johns-Manville
Ventsulation ASBESTOS FELT

This specification is for use over any type of structural deck (without insulation) either nailable or non-nailable and covers installation of Ventsulation Felt only. The balance of the built-up roofing membrane and flashings shall be installed according to their respective specifications. The Ventsulation System of venting at roof edges and parapets shall be followed.

In no case should the Ventsulation Felt be solidly or strip mopped. If nailing is not possible, these felts shall be spot mopped only.

Note: All information contained in "General Instructions" in the current Specification Manual for Johns-Manville Built-Up Roofs shall be considered part of this specification.

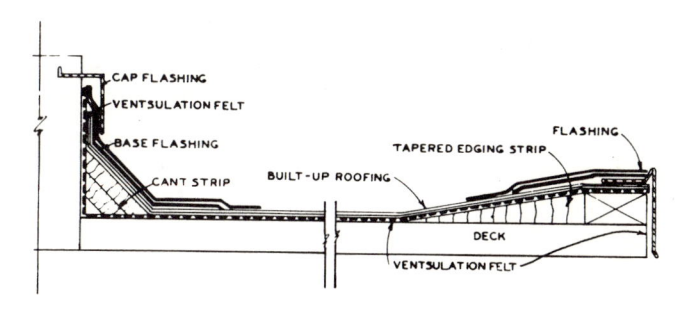

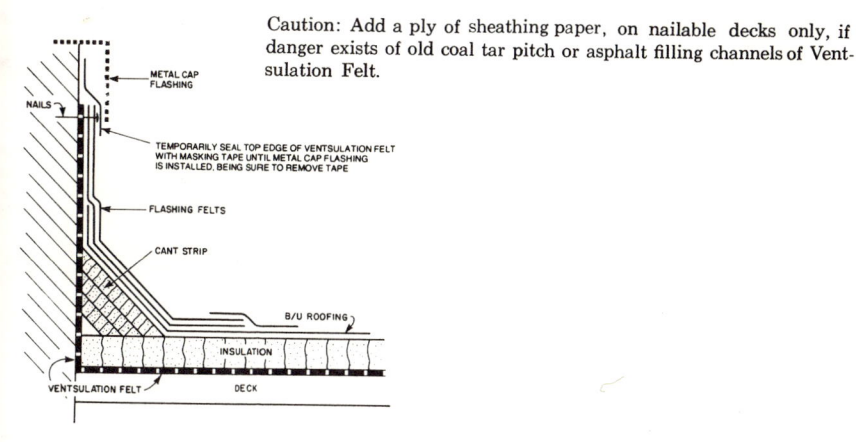

Caution: Add a ply of sheathing paper, on nailable decks only, if danger exists of old coal tar pitch or asphalt filling channels of Ventsulation Felt.

Application of Roofing

First: Apply one layer of Ventsulation Felt mineral surface down, lapping each felt the 1" selvage. End joints to be lapped 4".

Second: At roof edges carry the Ventsulation Felt up the tapered edging strip to the roof edge in such manner that venting can be accomplished behind the metal edging strip.

Third: At parapet walls carry the Ventsulation Felt under the cant strip and up the wall a minimum of 10". Secure sufficiently to hold in place until cant and flashing are installed. Cant strip, built-up roofing and base flashing are then installed in the usual manner. Do not seal the top of the Ventsulation Felt. Metal cap flashing is installed immediately so that venting space is left between it and the base flashing.

Fourth: If the deck is nailable, nail the laps at 9" centers and down the longitudinal center of each felt nail two rows of nails with the rows spaced approximately 11" apart, and nails staggered on approximately 18" centers. Use nails or fasteners appropriate to the type of deck.

If the deck is non-nailable the felt shall be secured to the deck with spots of hot J-M 190 or 220 Asphalt spaced on 24" centers.

Fifth: If the Ventsulation Felt is nailed apply either a smooth or gravel surfaced roof over the felt depending on the roofing specification selected. If the Ventsulation Felt is spot mopped only apply only a gravel surface roof over the felt.

Nailing

All nails or other fasteners are to be driven through tin caps unless the nail or fastener has an integral flat cap no less than 1" across.

BUILT-UP ROOFS by Johns-Manville / P.O. Box 5108, Denver, Colorado 80217

Specification **No. VS-2** *for use with*

APPROVED INSULATION

Johns-Manville
Ventsulation ASBESTOS FELT

This specification is for use over any type of structural deck and covers installation of Ventsulation Felt and Insulation only. The balance of the built-up roofing membrane and flashings shall be installed according to their respective specifications.

The Ventsulation Felt used with the insulation must be vented following the Ventsulation System of venting roof edges and parapets.

In no case should the Ventsulation felt be solidly or strip mopped. If nailing is not possible these felts shall be spot mopped only.

Note: All information contained in "General Instructions" in the current Specification Manual for Johns-Manville Built-Up Roofs shall be considered part of this specification.

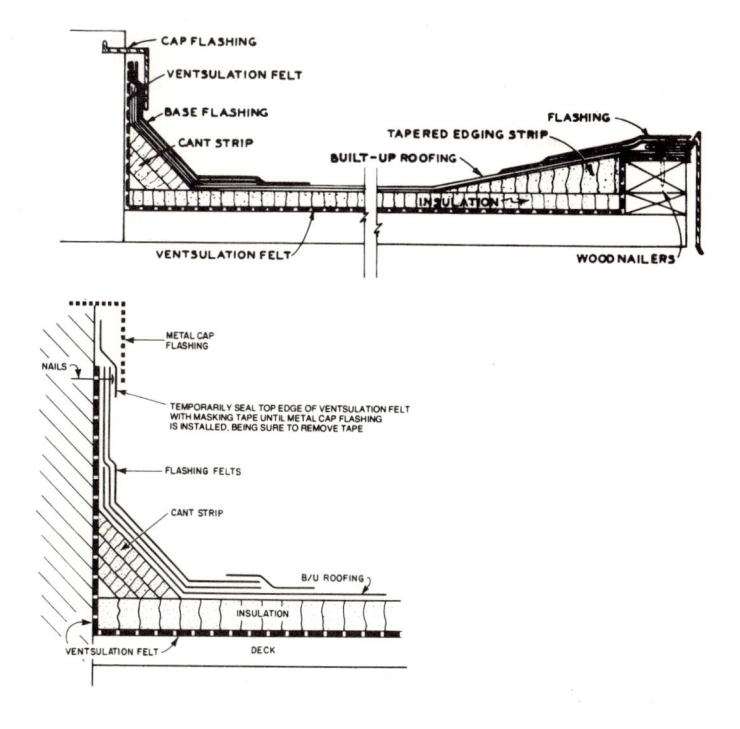

Application of Roofing

First: Apply one layer of Ventsulation Felt mineral surface down, lapping each felt the 1" selvage. End joints to be lapped 4".

Second: At roof edges carry the Ventsulation Felt up the tapered edging strip to the roof edge in such manner that venting can be accomplished behind the metal edging strip. As an alternate, the Ventsulation Felt can be carried out straight to the roof edge, prior to the installation of the wood nailer.

Third: At parapet walls carry the Ventsulation Felt under the cant strip and up the wall a minimum of 10". Secure sufficiently to hold in place until cant and flashings are installed. Cant strip, built-up roofing, and base flashing are then installed in the usual manner. Do not seal the top of the Ventsulation Felt. Metal cap flashing is installed immediately so that venting space is left between it and the base flashing.

Fourth: If the deck is nailable, nail the laps at 9" centers and down the longitudinal center of each felt, nail two rows of nails with the rows spaced approximately 11" apart and nails staggered on approximately 18" centers. Use nails or fasteners appropriate to the type of deck. If the deck is non-nailable, the Ventsulation Felt shall be secured to the deck with spots of hot J-M 190 or 220 Asphalt spaced on 24" centers.

Both side & end laps of Ventsulation Felt must be sealed with hot asphalt to be an effective vapor retarder under the insulation.

Fifth: Apply the units of insulation with the long joints continuous. Short joints shall be broken. Mop the full width under each unit of insulation. If the insulation is applied in more than one layer, all joints shall be broken between layers. The insulation shall not be left exposed to the weather. No more insulation shall be applied than can be completely covered with the roofing felts on the same day.

Sixth: Apply either a smooth or gravel surfaced roof over the insulation depending on the roofing specification selected.

Nailing

All nails or other fasteners are to be driven through tin caps unless the nail or other fastener has an integral flat cap no less than 1" across.

BUILT-UP ROOFS by Johns-Manville / P.O. Box 5108, Denver, Colorado 80217

Johns-Manville
Flashing Instructions

Inspection
All surfaces to be flashed shall be inspected before work is started since, in large measure, the success of a flashing depends on a properly constructed base.

Masonry construction
Walls shall be built with hard burned brick or sound re-inforced concrete. Common faults encountered are soft and scaling brick or concrete; poor mortar or faulty pointing of joints; and, broken copings and faulty pointing of joints between copings.

Also, walls of ordinary hollow tile, concrete blocks or other materials which in themselves are not waterproof, shall not be accepted as suitable to receive flashings unless they are properly waterproofed.

On all masonry surfaces which are to receive Asbestile or bitumen, the surface shall be primed with concrete primer regardless of the fact that these materials at times will adhere temporarily without the use of a primer.

Frame Construction
Frame walls shall not be accepted to receive flashing unless suitable solid backing for the flashing is provided. Flashing shall be installed so that the wall sheathing will finish over the flashing. In the case of stucco, suitable stops shall be provided to prevent the breaking away of the stucco over the top of the flashing.

Cants
Prior to application of flashings, a cant strip shall be installed to modify the angle between the roof deck and any vertical element of a structure.

Nailing Strips and Stucco Stops
As called for in flashing specifications, these elements shall be suitably installed by other than the roofing contractor.

Roof Edging Nailing Strips
In connection with non-nailable roof decks, wood nailing strips must be provided by other than the roofing contractor, to receive the flanges of metal edgings which must be secured by nailing.

Roofing Felts
All felts comprising the built-up roof shall be laid before the flashing is applied and shall be turned up as called for in the specification, but not cemented to, all surfaces to be flashed. Roofing felts shall not be carried up a wall to act as a base flashing.

Johns-Manville
Asbestile FLASHINGS
... made better with asbestos

More than any other place on the roof a leak is apt to develop at the intersection of the roof deck and a vertical surface. To give the required protection at such points, Johns-Manville has developed Asbestile Flashings.

These flashings take their name from Asbestile, a heavy bodied, plastic cement composed of asbestos fiber, asphalt and other mineral ingredients. When it sets, the Asbestile almost becomes an integral part of the wall itself.

To meet different conditions, Johns-Manville offers the five-course Super A Asbestile Flashings and the three-course Asbestile Flashings.

The five-course Asbestile Flashing is made up of a troweled layer of Asbestile, a layer of asbestos felt, another troweling of Asbestile, another layer of asbestos felt and a final troweling of Asbestile.

The three-course Asbestile Flashing is made up of a troweled layer of Asbestile, a layer of asbestos felt and a final troweling of Asbestile.

Asbestile Flashings provide thorough water-tightness and maximum protection at those critical points where a roof deck is intersected by a vertical surface.

Specification **No. FE-1** *for application over*

MASONRY CONSTRUCTION WITH NAILING FACILITIES

Johns-Manville

Asbestile SUPER "A" FLASHING

Base and Cap—with metal cap—for use with bituminous built-up roofs

All information in "Flashing Instructions", Specification Manual for Johns-Manville Built-Up Roofs, shall be considered part of this specification.

Preparation for Flashing — If a two piece metal cap flashing is installed prior to application of base flashing the flange of the metal cap flashing shall be removed to allow for application of J-M Asbestile Base and Cap Flashing.

Primer — Coat with concrete primer the masonry surface over which Asbestile Base and Cap Flashings are to be applied and the vertical surface to the point where the metal flashing enters the wall.

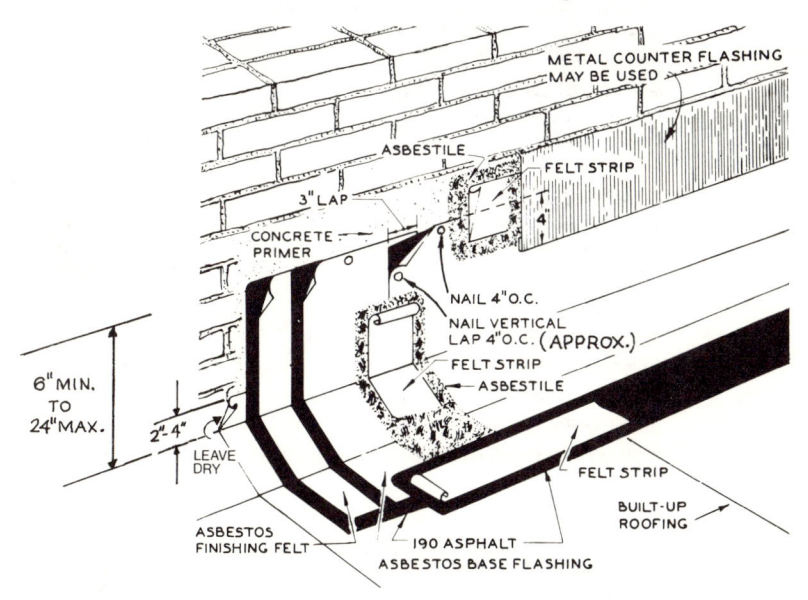

Application of Base Flashing

Roofing felts must extend at least 2" to 4" above the top of the cant strip and be left dry to provide a plane of slippage behind the base flashing. All base flashing materials shall extend not less than 6" high on the vertical surfaces and not less than 4" on the roof. Such dimensions shall be measured from the top and bottom edges of the cants.

This specification is eligible for a 20-Year Endorsement only when, in the opinion of an authorized J-M Representative, all conditions listed in "General Instructions" of this Specification Manual have been met.

Starting 3" below the point where the metal flashing enters the wall, mop with hot J-M 190 Asphalt the area to receive the backer felt of J-M Asbestile Finishing Felt. Press it into place, lapping the ends of the felt 3". Mop with hot J-M 190 Asphalt the surface of the backer felt just applied. Then, mop the back surface of the J-M Base Flashing and press it into place. Nail thru tin caps the flashing at 4" centers along the top edge using 1" long concrete nails. Lap the ends of the flashing 3" and nail the lap on 4" centers (approx.) vertically. Cover the lap with Three Course Asbestile (a 4" wide strip of J-M Asbestos Finishing Felt embedded in and troweled over with layers of Asbestile 1/8" thick).

Note: Mineral Surfaced roofing may be substituted for the Asbestos Base Flashing in Region 3.

Cover the roof edge of the base flashing with a 4" wide strip of J-M Asbestos Finishing Felt embedded in and coated with the bitumen.

Application of Asbestile Cap Flashing

Centered on the top line of the base flashing and completely covering all nails, trowel a layer of Asbestile 1/8" thick and 5" wide. Into this Asbestile, embed a 4" wide strip of J-M Asbestos Finishing Felt. Then, trowel another layer of J-M Asbestile 1/8" thick and 5" wide bringing the Asbestile to a feather edge.

Metal Cap Flashing

The metal flashing if used shall be installed according to the manufacturer's specifications and shall be applied to overlap the base flashing at least 4".

Surfacing

The exposed area of the base flashing may be finished with any J-M Approved Roof Coating, except Asphalt.

Materials

Base Flashing: Approximately 80 lin ft of completed flashing can be constructed with the following materials:

Concrete Primer	1 gal
J-M Asbestos Finishing Felt for backer felt	1 sq
J-M Asbestos Base Flashing	1 sq
J-M Asbestos Finishing Felt 4" wide	1 roll
J-M 190 Asphalt	80 lb
Asbestile	1/2 gal.

Asbestile Cap Flashing: Three-Course Asbestile counter flashing. Approximately 100 lin. ft of completed flashing can be constructed with the following:

Concrete Primer	1 gal
J-M Asbestos Finishing Felt, 4"	1 roll
J-M Asbestile	5 gals.

BUILT-UP ROOFS *by Johns-Manville/Greenwood Plaza, Denver, Colorado 80217*

Specification **No. FE-2** *for application over*

MASONRY CONSTRUCTION

Johns-Manville
Asbestile SUPER "A" CAP FLASHING
For use with bituminous built-up roofs

All information in "Flashing Instructions", Specification Manual for Johns-Manville Built-Up Roofs, shall be considered part of this specification.

Preparation for Flashing — All base flashing materials must be in place and shall extend not less than 6" high on the vertical surfaces and not less than 4" on the roof. Such dimensions shall be measured from the top and bottom edges of the cant.

Primer — Coat with concrete primer the masonry surface over which cap flashing is to be applied.

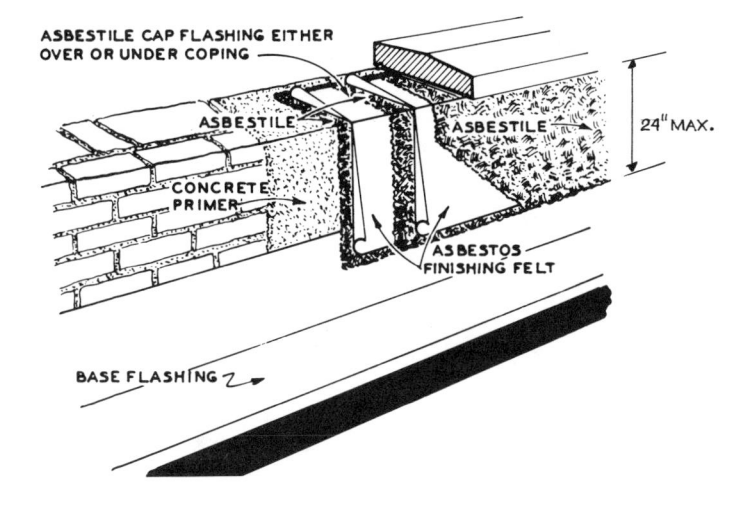

This specification is eligible for a 20-Year Endorsement only when, in the opinion of an authorized J-M Representative, all conditions listed in "General Instructions" of this Specification Manual have been met.

Application of Cap Flashing

This cap flashing can be applied either over or under coping. Starting 2" from the outer top edge of the parapet wall, trowel a 1/8" thick layer of J-M Asbestile over the top and down the inside face to a line 4" below the upper edge of the base flashing. Into this Asbestile, embed a layer of J-M Asbestos Finishing Felt. Lap the felt edges 3" and seal with Asbestile.

Similarly, trowel a second layer of Asbestile 1/8" thick and embed a second layer of J-M Asbestos Finishing Felt.

Trowel a final layer of J-M Asbestile 1/8" thick bringing it to a feather edge.

Surfacing

The exposed area of the cap flashing may be finished with any J-M Approved Roof Coating, except Asphalt.

Materials

Cap Flashing: Five-Course Asbestile cap flashing. Approximately 100 sq ft of completed flashing can be constructed with the following:

Concrete Primer	1 gal
J-M Asbestos Finishing Felt (Perforated)	2 sq
J-M Asbestile	20 gals*

*If surface receiving Asbestile is unusually rough, up to 3 gals additional may be required.

BUILT-UP ROOFS *by Johns-Manville/Greenwood Plaza, Denver, Colorado 80217*

*Specification **No. FE-3** for application over*
MASONRY WALLS WITH NO NAILING FACILITIES

Johns-Manville
Asbestile FIVE-COURSE FLASHING
Full height of parapet and over top—for use with bituminous built-up roofs

All information in "Flashing Instructions", Specification Manual for Johns-Manville Built-Up Roofs, shall be considered part of this specification.

Primer — Coat with concrete primer the masonry surface over which the flashing is to be applied.

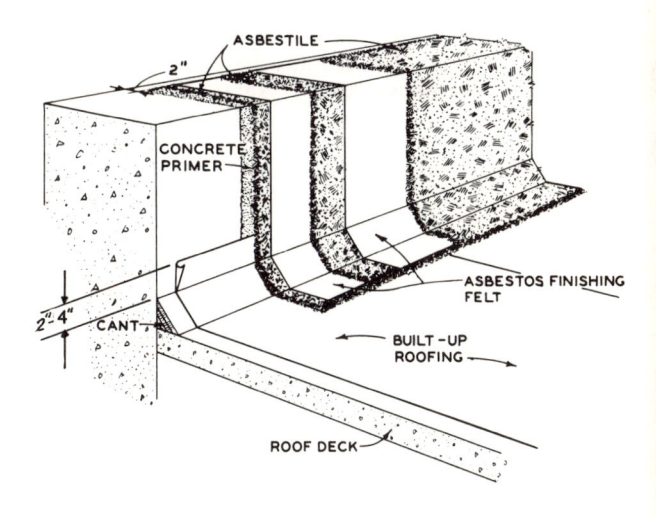

Roofing felts must extend at least 2" to 4" above the top of the cant strip and be left dry to provide a plane of slippage behind the base flashing.

This specification is eligible for a 20-Year Endorsement only when, in the opinion of an authorized J-M Representative, all conditions listed in "General Instructions" of this Specification Manual have been met.

Application of Flashing

Starting 2" from the outer edge of the parapet, trowel a 1/8" thick layer of J-M Asbestile over the top, down the inside face, across the cant and extend it 4" out on the roof. Into this Asbestile, embed a layer of J-M Asbestos Finishing Felt. Lap the felt ends 3" and seal with J-M Asbestile.

Similarly, trowel a second layer of J-M Asbestile 1/8" thick and embed a second layer of J-M Asbestos Finishing Felt.

Trowel a final layer of Asbestile 1/8" thick bringing it to a feather edge.

Surfacing

The flashing may be finished with any J-M Approved Roof Coating, except Asphalt.

Materials

Five Course Asbestile Flashing: Approximately 100 sq ft of completed flashing can be constructed with the following:

Concrete Primer	1 gal
J-M Asbestos Finishing Felt	2 sq
J-M Asbestile	20 gals*

*If surface receiving Asbestile is unusually rough, up to 3 gals additional may be required.

BUILT-UP ROOFS by Johns-Manville/Greenwood Plaza, Denver, Colorado 80217

Specification **No. FE-4** *for application over*
MASONRY WALLS WITH NO NAILING FACILITIES

Johns-Manville
Asbestile FIVE-COURSE FLASHING
At least 10" up the wall—for use with bituminous built-up roofs

All information in "Flashing Instructions", Specification Manual for Johns-Manville Built-Up Roofs, shall be considered part of this specification.

Primer — Coat with concrete primer the masonry surface over which the flashing is to be applied.

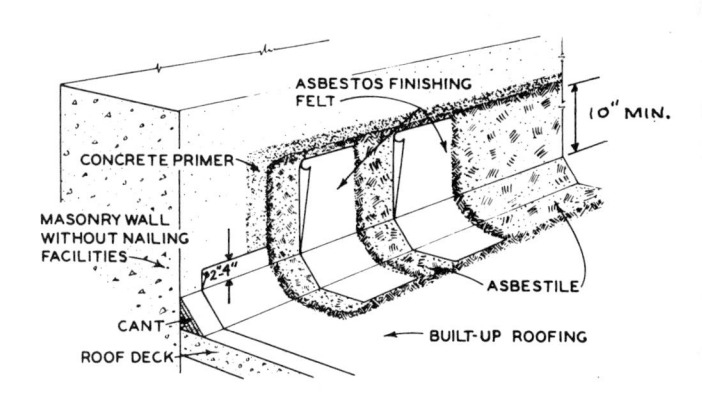

Roofing felts must extend at least 2" to 4" above the top of the cant strip and be left dry to provide a plane of slippage behind the base flashing.

This specification is eligible for a 20-Year Endorsement only when, in the opinion of an authorized J-M Representative, all conditions listed in "General Instructions" of this Specification Manual have been met.

Application of Flashing

Starting at a line at least 10″ above the cant, trowel a 1/8″ thick layer of J-M Asbestile down the inside face, across the cant and extend it 4″ out on the roof. Into this Asbestile, embed a layer of J-M Asbestos Finishing Felt. Lap the felt ends 3″ and seal with Asbestile.

Similarly, trowel a second layer of Asbestile 1/8″ thick and embed a second layer of J-M Asbestos Finishing Felt.

Trowel a final layer of J-M Asbestile 1/8″ thick bringing it to a feather edge.

Surfacing

The flashing may be finished with any J-M Approved Roof Coating, except Asphalt.

Materials

Five-Course Asbestile Flashing — Approximately 100 sq ft of completed flashing can be constructed with the following:

Concrete Primer	1 gal
J-M Asbestos Finishing Felt	2 sq
J-M Asbestile	20 gals*

*If surface receiving Asbestile is unusually rough, up to 3 gals additional may be required.

BUILT-UP ROOFS by Johns-Manville/Greenwood Plaza, Denver, Colorado 80217

Specification **No. FE-10** *for application over*

MASONRY WALLS WITH NAILING FACILITIES

Johns-Manville

Asbestile STANDARD FLASHING

Base and Cap—with metal cap—for use with bituminous built-up roofs

All information in "Flashing Instructions", Specification Manual for Johns-Manville Built-Up Roofs, shall be considered part of this specification.

Preparation for Flashing — All base flashing materials must be nailed into the wall using 1" long concrete nails. If a wall does not permit such nailing, suitable wood strip must be cast into the wall by other than the roofing contractor.

If a two piece metal cap flashing is installed prior to the application of the base flashing the flange of the metal cap flashing shall be removed to allow for application of the base flashing.

Primer — Coat with concrete primer the brick surface over which J-M Asbestile Base and Cap Flashings are to be applied and the vertical surface to the point where the metal flashing enters the wall.

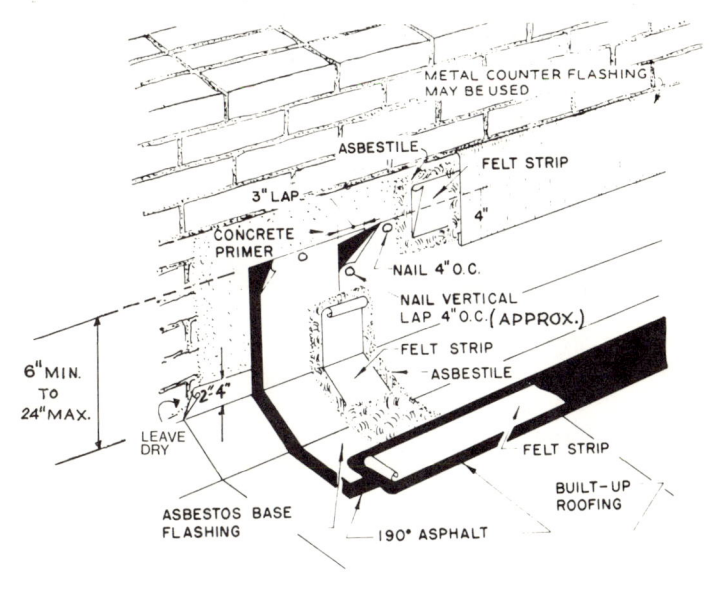

This specification is eligible for a 15-Year Endorsement only when, in the opinion of an authorized J-M Representative, all conditions listed in "General Instructions" of this Specification Manual have been met.

Application of Base Flashing

Roofing felts must extend at least 2" to 4" above the top of the cant strip and be left dry to provide a plane of slippage behind the base flashing.

Measured from the top of the cant and up the inside face of the parapet wall, the minimum height for this base and cap flashing shall be not less than 6" high on the vertical surfaces and not less than 4" on the roof. Such dimensions shall be measured from the top and bottom edges of the cants.

Mop the area to receive the J-M Base Flashing with hot J-M 190 Asphalt. Then mop the back surface of the base flashing and press it into place. Nail thru tin caps the flashing at 4" centers along the top edge. Lap the ends of the flashing 3" and nail the lap on 4" (approx.) centers vertically. Cover the lap with Three-Course Asbestile (a 4" wide strip of J-M Asbestos Finishing Felt embedded in and troweled over the layers of J-M Asbestile 1/8" thick).

Cover the roof edge of the base flashing with a 4" wide strip of J-M Asbestos Finishing Felt embedded in and coated with the bitumen.

Application of Asbestile Cap Flashing

Centered on the top line of the base flashing and completely covering all nails, trowel a layer of J-M Asbestile 1/8" thick and 5" wide. Into this Asbestile, embed a 4" wide strip of J-M Asbestos Finishing Felt. Then trowel another layer of J-M Asbestile 1/8" thick and 5" wide bringing the Asbestile to a feather edge.

Metal Cap Flashing

The metal cap flashing if used shall be installed in accordance with the manufacturer's specifications and shall be installed to overlap the base flashing at least 4".

Surfacing

The exposed area of the base flashing may be finished with any J-M Approved Roof Coating, except Asphalt.

Materials

Base and Cap Flashing — Approximately 80 lin ft of completed flashing can be constructed with the following materials:

Concrete Primer	1 gal
J-M Asbestos Base Flashing	1 sq
J-M 15 lb Asbestos Finishing Felt, 4" wide	2/3 roll
J-M 190 Asphalt	50 lb
J-M Asbestile	5 gals

BUILT-UP ROOFS by Johns-Manville/Greenwood Plaza, Denver, Colorado 80217

Specification **No. FE-20** *for application over*

FRAME CONSTRUCTION

Johns-Manville
Asbestile FLASHING
Base Flashing under wood siding—for use over bituminous built-up roofs

All information in "Flashing Instructions", Specification Manual for Johns-Manville Built-Up Roofs, shall be considered part of this specification.

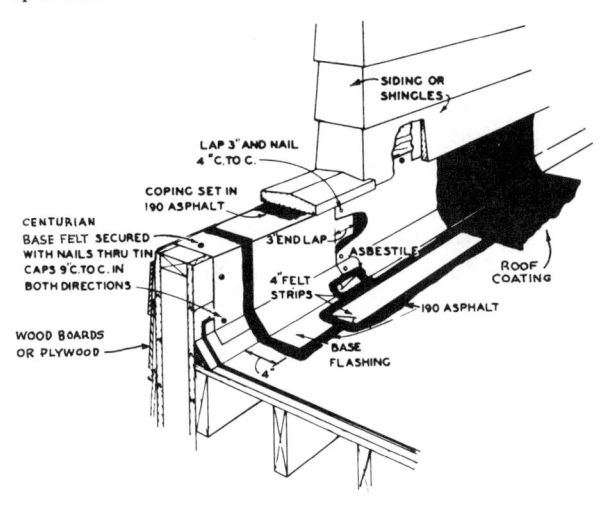

This specification is eligible for a 20-Year Endorsement only when, in the opinion of an authorized J-M Representative, all conditions listed in "General Instructions" of this Specification Manual have been met.

Application of Flashing over Parapet Wall

Starting at the outer edge of the parapet wall, apply a layer of J-M Centurian Base Felt extending it over the top and down the inside face to a line along the middle of the cant. Lap the felt ends 3" and nail the entire layer on 9" centers in both directions.

Apply a layer of J-M Asbestos Base Flashing starting at the outer edge of the parapet wall and extending across the top, down the inside face, over the cant and onto the roof at least 4". The back surface of the base flashing shall be mopped with hot J-M 190 Asphalt before application and the flashing shall be imbedded in a mopping of hot J-M 190 Asphalt.

Lap the ends of the flashing 3", nail the laps on 4" centers vertically and cover the lap with a 4" wide strip of J-M Asbestos Finishing Felt embedded in J-M Asbestile.

Note: Mineral Surfaced roofing may be substituted for the Asbestos Base Flashing in Region 3.

Cover the roof edge of the flashing with a 4" strip of felt embedded in bitumen and finish the parapet wall with a suitable wood coping.

Surfacing

The exposed area of the flashing may be finished with any J-M Approved Roof Coating, except Asphalt.

Application of Flashings under wood finish

This application is the same as for "over parapet wall" except . . . The J-M Centurian Base Felt and the J-M Asbestos Base Flashing is applied to the vertical wood wall. The top edge of the J-M Asbestos Base Flashing is then nailed on 4" centers. This edge then is overlapped at least 2" by the clapboards or shingles of the finished wall.

BUILT-UP ROOFS by Johns-Manville/Greenwood Plaza, Denver, Colorado 80217

*Specification **No. FE-30** for*

Roof Edgings and Gravel Stops

For roofs without parapets or copings, a metal edging is usually employed to give a building a finished appearance. This edging covers the junction between roof and sidewalls, provides a decorative facia for the building and, when required, acts as a gravel stop.

Edging treatment flush with the roof deck is not recommended.

Prior to the application of any metal edging all felts of the built-up roofing membrane are to be carried up over the tapered edging strip and secured to the wood nailer.

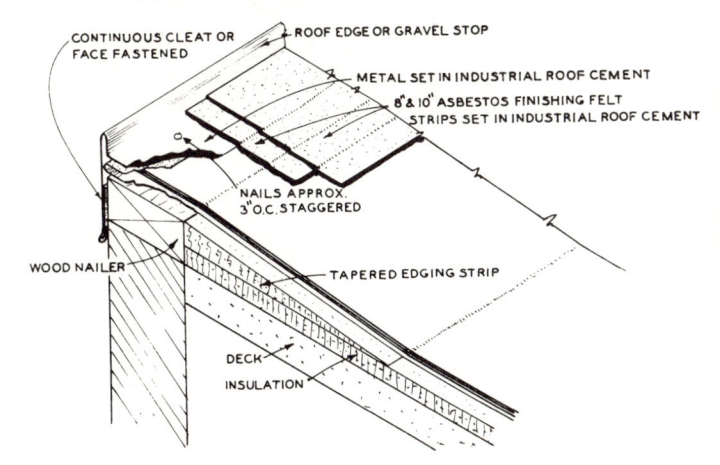

[Diagram labels: CONTINUOUS CLEAT OR FACE FASTENED; ROOF EDGE OR GRAVEL STOP; METAL SET IN INDUSTRIAL ROOF CEMENT; 8" & 10" ASBESTOS FINISHING FELT STRIPS SET IN INDUSTRIAL ROOF CEMENT; NAILS APPROX. 3" O.C. STAGGERED; WOOD NAILER; TAPERED EDGING STRIP; DECK; INSULATION]

The construction shown above should be used with light gauge metals such as copper, hot galvanized steel, or aluminum. This detail uses a tapered edging strip to raise the edging above the roof level. The lower ply of roofing felt should be extended and folded back over completing plys of roofing to form an envelope to reduce possibility of bitumen drip.

These metal edgings have roof flanges which are secured by nailing 3" O.C. Therefore, wood nailing strips must be provided by other than the roofing contractor and should be at least 2" wider than the nailing flange of the metal edging strip to provide adequate nailing. If roof insulation is used the wood nailer must be the same thickness as the insulation and the tapered edging strip.

After the metal edging is nailed in place, over the the roofing felts, the flanges are covered by two strips of J-M Asbestos Finishing Felt. The minimum width of the first strip is 8" wide and the second strip is 10" wide and both are to be embedded in J-M Industrial Roof Cement. If it is a smooth-surface roof, cover the strips with the same surfacing of bitumen used on the rest of the roof. If it is a gravel or slag roof employing organic felts, the asbestos finishing felts must be extended and carried completely down the tapered edging strip and out on to the roof completely covering the organic felts. Coat these felts with the same bitumen used for the flood coat.

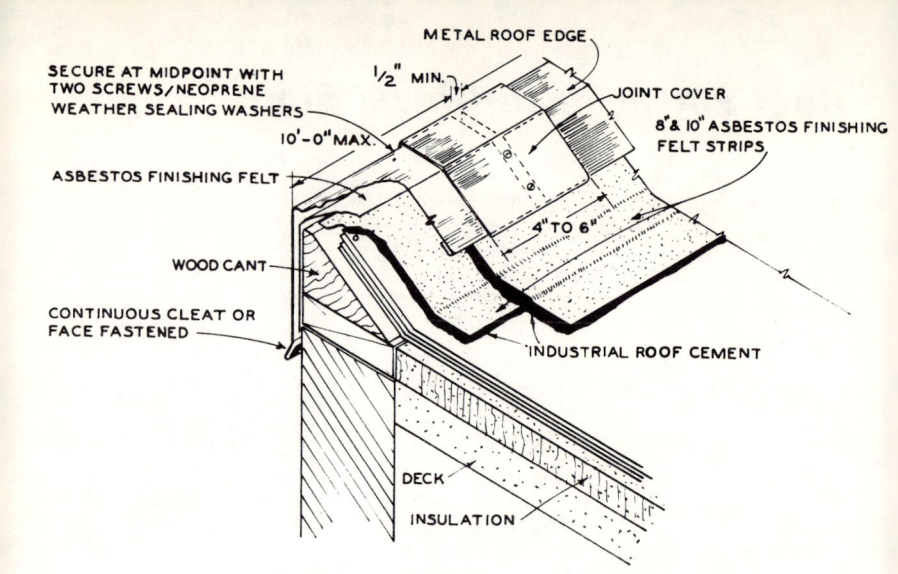

This detail shows another method of treating roof edges utilizing a wood cant strip to further raise the metal edging above the roof level. An additional strip of Asbestos Finishing is draped down over the cant to protect the exposed edge of the plys of roofing felts which have been carried up the cant strip. This also seals the flashing system until metal work is installed.

BUILT-UP ROOFS by Johns-Manville/Greenwood Plaza, Denver, Colorado 80217

Specification **No. FE-40**

Johns-Manville
Flashings for Roof Fixtures

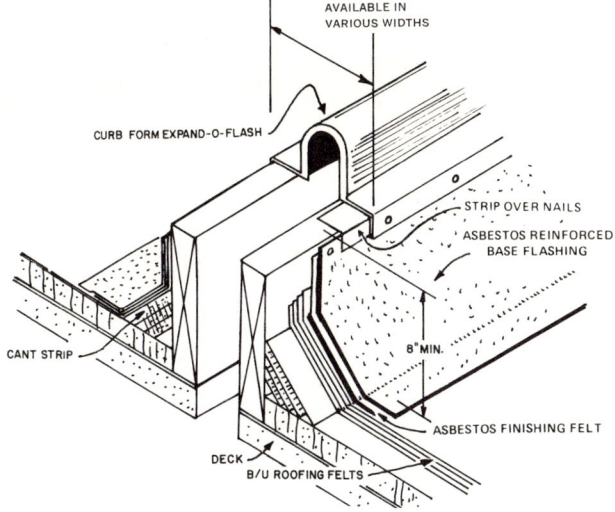

This type of expansion joint cover, curb mounted, allows for building movement in all directions.

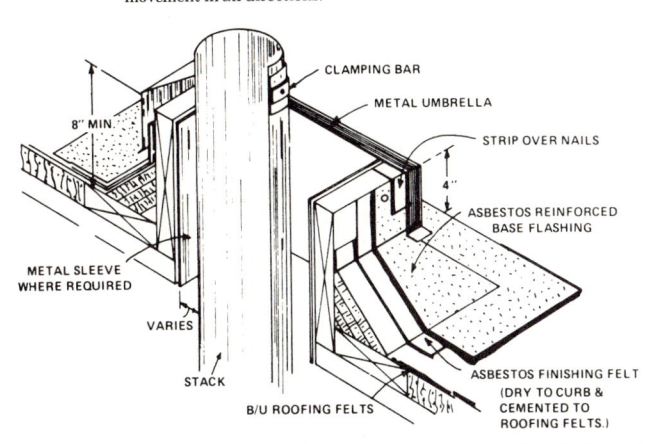

Curbs must be provided for any accessory passing through the roof deck such as a vent pipe or conduit. Curbs must also be provided for roof fixtures such as skylights or ventilating fans.

The detail shown allows the opening to be completed and roofing installed prior to placing any accessory.

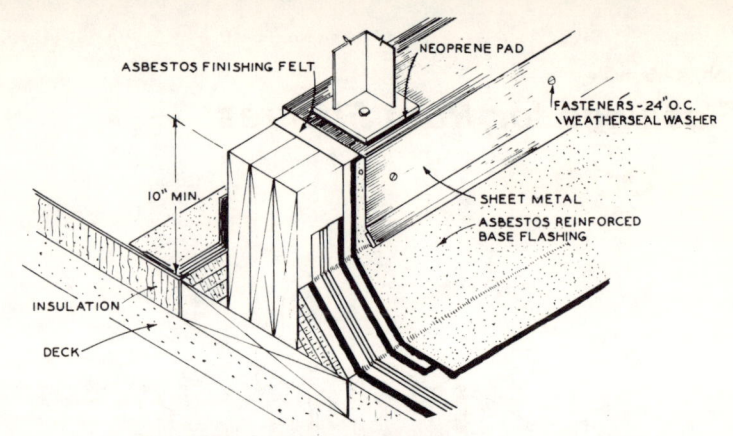

Equipment or Sign Support

By raising supports above roof level maintenance is simplified. For heavy loads supports should be continuous and located over structural members such as girders or columns.

Pitch pans or pockets are not recommended.

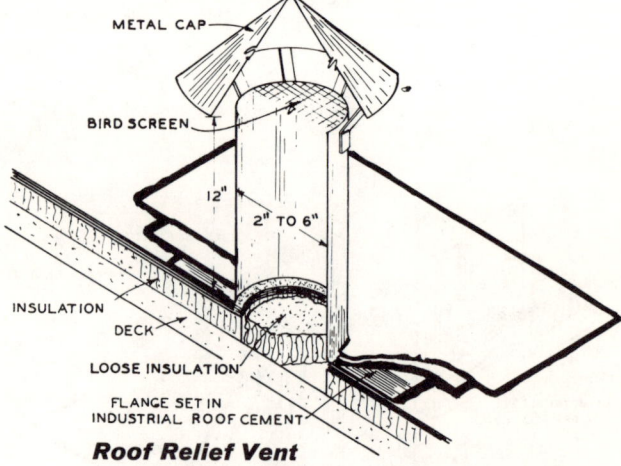

Roof Relief Vent

Roof vents are used to reduce moisture content of wet insulation or certain types of wet fill poured decks. They may also be used in conjunction with Ventsulation Felt when parapet or roof edge venting is not practical. Placement of vents depends on job conditions but there should be a minimum of one vent for each 10 squares of roof area.

Note: These flashing details follow current recommendations of the National Roofing Contractors Association and are used with their sanction.

BUILT-UP ROOFS by Johns-Manville/Greenwood Plaza, Denver, Colorado 80217

Re-roofing

Insulated decks of the pre-formed insulation or light-weight aggregate concrete type are likely to contain and retain considerable moisture if the membrane has been allowed to deteriorate. Because of this moisture in the deck precaution must be taken, when applying a new roof, not to seal the moisture in the assembly.

The first and best solution is to completely remove the old or damaged membrane, make necessary repairs to deck, and re-roof. However, in many instances where the insulation is still sound it is uneconomical to remove the old membrane; therefore, re-roofing can be accomplished by one of the methods on the following pages, depending on the type of deck to be re-roofed.

Specification **No. RR-1**

Johns-Manville
Re-roofing
Over insulation on nailable or non-nailable decks

If existing roof is gravel surfaced, all old gravel must be completely removed so that the old membrane is free of all gravel. Prior to application of Ventsulation Felt make random cuts or breaks in old membrane to permit escape of moisture from the insulation. The Ventsulation Felt used as the base felt must be vented following the Ventsulation System of venting at roof edges and parapets.

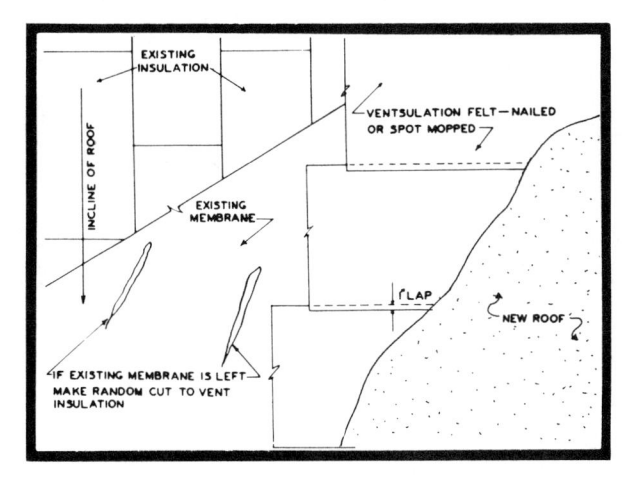

CAUTION: Add one ply of sheathing paper, on nailable decks only, if danger exists of old coal tar pitch or asphalt filling the channels of Ventsulation Felt.

Application of Roofing

First, over the old membrane, which has been cut to allow moisture to escape, apply one layer of Ventsulation Felt lapping each felt the 1" selvage. End joints to be lapped 4". If the deck is nailable, nail the laps at 9" centers and down the longitudinal center of each felt nail two rows of nails with the rows spaced approximately 11" apart and nails staggered on approximately 18" centers. Use nails or fasteners appropriate to the type of deck.

If the existing insulation is fiberboard and at least 1" thick, secure Ventsulation Felt using suitable fastener.

Use standard nailing pattern

If the deck is non-nailable the Ventsulation Felt shall be secured to the deck with spots of hot J-M 190 Asphalt spaced on 24" centers.

Second, apply either a smooth or gravel surfaced roof over the Ventsulation Felt. Where the Ventsulation Felt is spot-mopped but not nailed use only gravel-surface roof.

Nailing

All nails or other fasteners are to be driven through tin caps unless the nail or fastener has an integral flat cap no less than 1" across.

Parapet and Edging Details

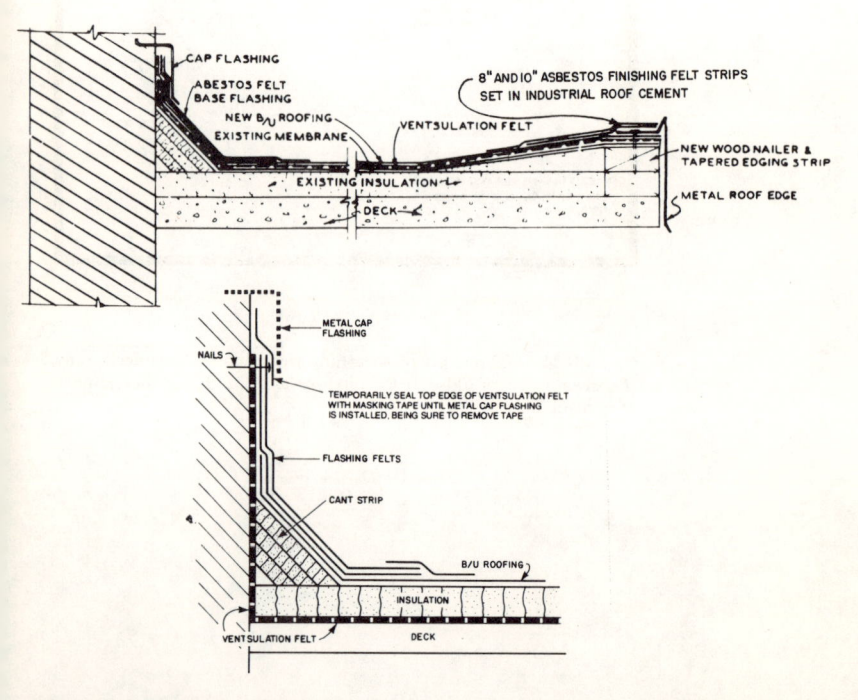

BUILT-UP ROOFS by Johns-Manville / P.O. Box 5108, Denver, Colorado 80217

Specification **No. RR-2**

Johns-Manville
Re-roofing
Over lightweight concrete aggregate decks

If existing roof is gravel surfaced, all old gravel must be completely removed so that the old membrane is free of all gravel. Prior to application of Ventsulation Felt make random cuts or breaks in old membrane to permit escape of moisture from the deck. The Ventsulation Felt used as the base felt must be vented following the Ventsulation System of venting at roof edges and parapets.

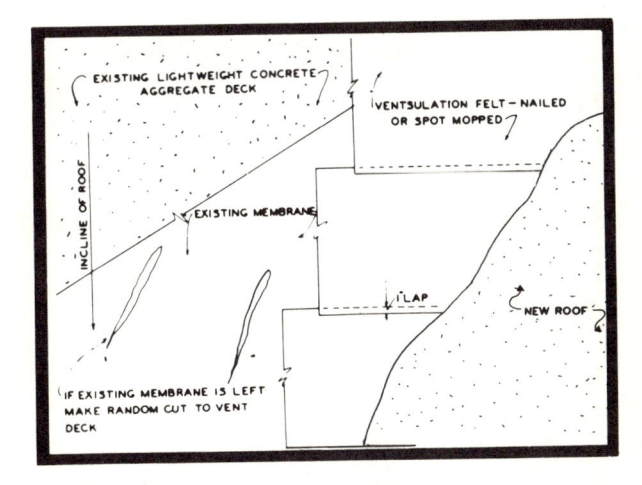

CAUTION: Add one ply of sheathing paper, on nailable decks only, if danger exists of old coal tar pitch or asphalt filling the channels of Ventsulation Felt.

Application of Roofing

First, over the old membrane, which has been cut to allow moisture to escape, apply one layer of Ventsulation Felt lapping each felt the 1" selvage. End joints to be lapped 4". If the deck is nailable, nail the laps at 9" centers and down the longitudinal center of each felt nail two rows of nails with the rows spaced approximately 11" apart and nails staggered on approximately 18" centers. Use nails or fasteners appropriate to the type of deck.

If the deck is non-nailable the Ventsulation Felt shall be secured to the deck with spots of hot J-M 190 Asphalt spaced on 24" centers.

Second: On a nailable deck apply either a smooth surface or gravel surface roof over the Ventsulation Felt.

On a non-nailable deck apply only a gravel surface roof over the Ventsulation Felt.

Nailing

All nails or other fasteners are to be driven through tin caps unless the nail or fastener has an integral flat cap no less than 1" across.

Parapet and Edging Details

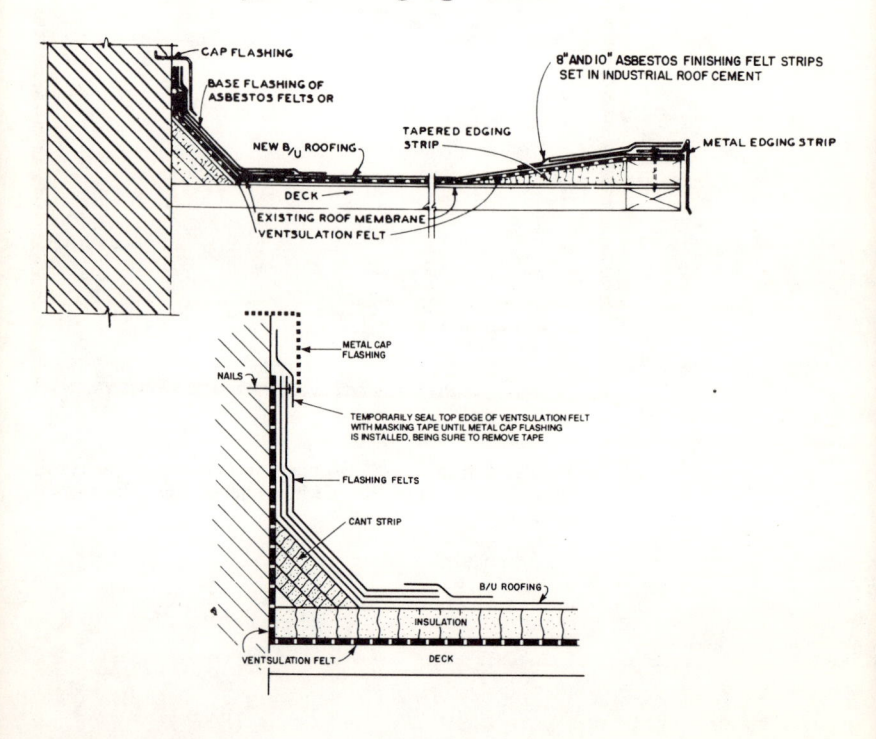

BUILT-UP ROOFS by Johns-Manville / P.O. Box 5108, Denver, Colorado 80217

Specification **No. RR-3**

Johns-Manville
Re-roofing
Over uninsulated decks — Nailable or Non-Nailable

If existing roof is gravel surfaced all old gravel must be completely removed so that the old membrane is free of all gravel. The Ventsulation must be vented at roof edges and parapets.

In no case should the base felt be solidly or strip mopped. If nailing is not possible these felts shall be spot mopped only.

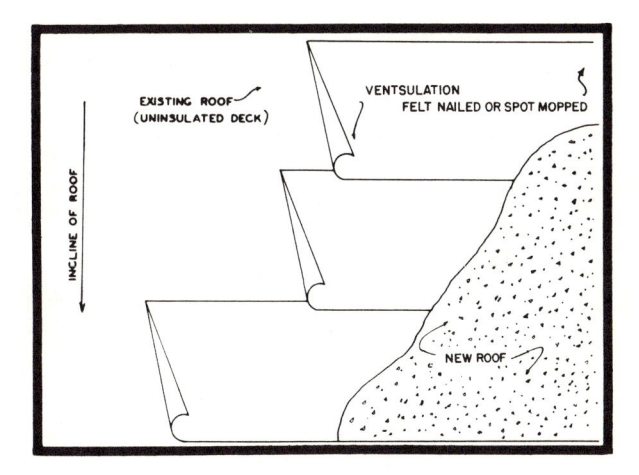

CAUTION: Add one ply of sheathing paper, on nailable decks only, if danger exists of old coal tar pitch or asphalt filling the channels of Ventsulation Felt.

Application of Roofing

First: Apply one layer of the base felt lapping each felt a minimum of 2". Ends shall be lapped 4". Ventsulation Felt must be used as the base felt if the deck or old membrane is suspected of being wet.

Second: If the deck is nailable, nail the laps at 9" centers and down the longitudinal center of each felt nail two rows of nails with the rows spaced approximately 11" apart, and nails staggered on approximately 18" centers. Use nails or fasteners appropriate to the type of deck.

If the deck is non-nailable the base felt shall be secured to the deck with spots of hot J-M 190 Asphalt spaced on 24" centers.

Third: Apply either a smooth or gravel surfaced roof over the base felt, if it is nailed. Where the base is spot mopped only use a gravel surface specification.

Nailing

All nails or other fasteners are to be driven through tin caps unless the nail or fastener has an integral flat cap no less than 1" across.

Parapet and Edging Details

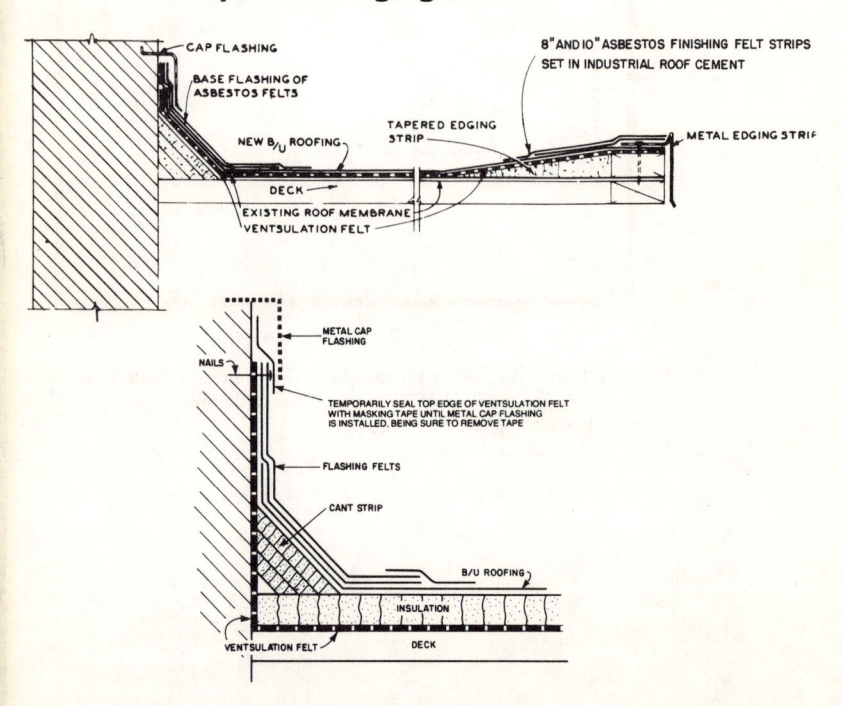

BUILT-UP ROOFS by Johns-Manville / P.O. Box 5108, Denver, Colorado 80217

Specification **No. RR-4**

Johns-Manville

Re-roofing *Over insulation over existing membrane*

If existing roof is gravel surfaced all old gravel must be completely removed so that the old membrane is completely free of all gravel.

The Ventsulation Felt used under the new insulation must be vented following the Ventsulation System of venting at roof edges and parapets.

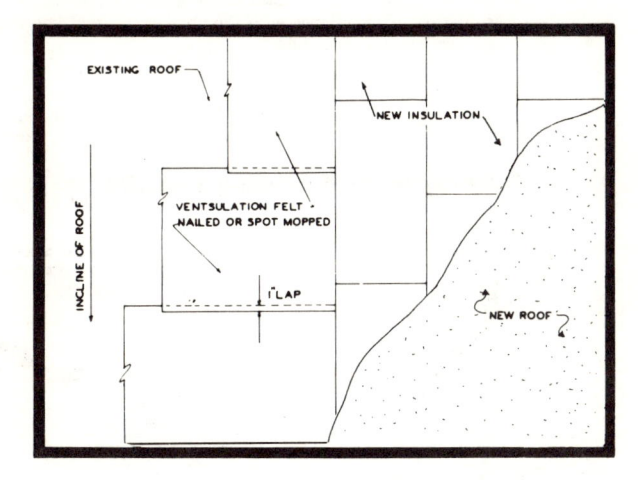

CAUTION: Add one ply of sheathing paper, on nailable decks only, if danger exists of old coal tar pitch or asphalt filling the channels of Ventsulation Felt.

Application of Roofing

First: Over the old membrane apply one layer of Ventsulation Felt lapping each felt the 1" selvage. End joints to be lapped 4". If the deck is nailable, nail the laps at 9" centers and down the longitudinal center of each felt nail two rows of nails with the rows spaced approximately 11" apart and nails staggered on approximately 18" centers. Use nails or fasteners appropriate to the type of deck.

If the deck is non-nailable the Ventsulation Felt shall be secured to the deck with spots of hot J-M 190 Asphalt spaced on 24" centers.

Both side & end laps of Ventsulation Felt must be sealed with hot asphalt to be an effective vapor barrier under the insulation.

Second: Apply the insulation by mopping with hot asphalt in accordance with the instructions of the manufacturer of the insulation.

Third: Apply either a smooth or gravel surfaced roof over the insulation in accordance with the appropriate roof specifications for application over insulation.

Nailing

All nails or other fasteners are to be driven through tin caps unless the nail or fastener has an integral flat cap no less than 1" across.

Parapet and Edging Details

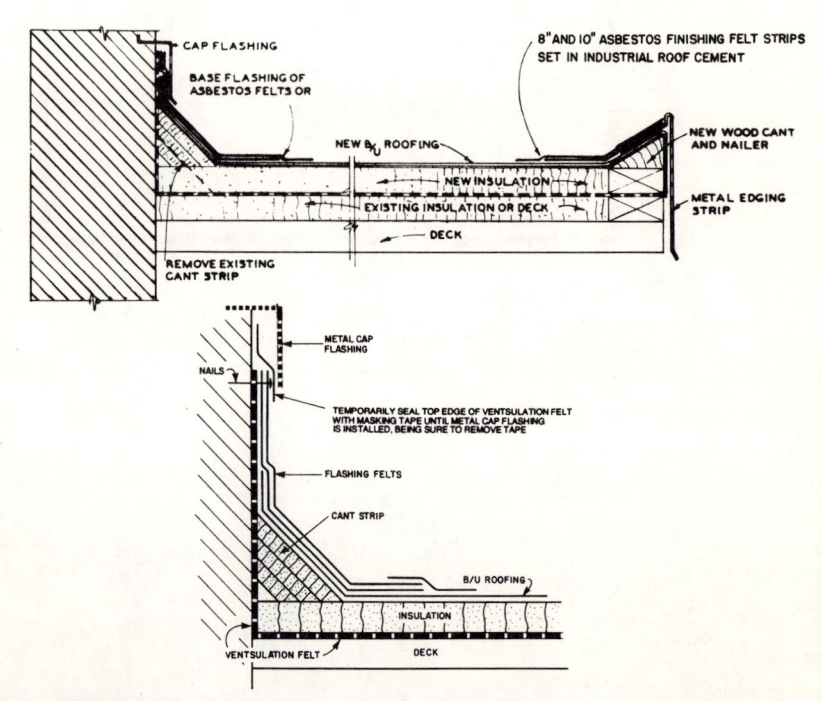

BUILT-UP ROOFS by Johns-Manville / P.O. Box 5108, Denver, Colorado 80217

Johns-Manville
Temporary Roofing

Temporary Roofing is a minimal membrane placed to permit occupancy during completion of construction. Temporary roofing is desirable when the building must be covered during extremely cold or inclement weather, or when it is necessary to close in the building before the trades have completed work at the roof level or above which requires use of the roof deck as a working platform.

Temporary Roofing should never be used as a part of the permanent roof. If the deck is steel and requires an insulation layer, the insulation and temporary roof should be completely removed prior to application of the permanent roof.

A temporary covering should not be used as part of the permanent roof for several reasons. It is applied as an emergency action with full knowledge that it will be abused either by the weather or by construction activity or both. If the weather is extremely cold or inclement the application conditions will probably cause it to be damaged from underside moisture, moisture between plies and laps, or from standing water freezing and thawing. If construction activity continues over it, it will become worn and covered with dirt, and probably punctured and torn in numerous places. Asphalt does not bond well to dirty surfaces. Voids between felts would result. Worn and aged felt absorbs and holds moisture readily. This moisture would start blister growth in a membrane placed over it. Use of such a damaged membrane as part of the permanent roof would probably result in a premature failure of the entire membrane.

Over Nailable Decks

First: Apply a sheathing paper over the deck, nailing sufficiently to hold in place.

Second: Apply a coated or inorganic base felt, nailing along the top edge, and lapping the next course 4" over the first, covering all nails, and cementing the lap and end laps with hot asphalt or cold application cement.

Third: (Optional) For extra protection mop on a surfacing of 190° asphalt.

Fourth: When the permanent roof is to be applied, tear off all the temporary covering and discard it. Prepare the deck and apply permanent roof.

OR

If insulation is to be used, clean the temporary covering of all debris and foreign matter, repair breaks where necessary, nail on at least a ply of #15 felt and mop the insulation to it as specified.

Over Non-Nailable Decks (except steel)

First: Prime the deck if priming is required.

Second: Apply one ply of coated base felt spot mopping it with spots about 12" in diameter and about 18" apart. Lap the courses 4" solidly cementing the side and end laps.

Third: (Optional) For extra protection mop on a surfacing of 190° asphalt.

Fourth: When the permanent roof is to be applied, tear off all the temporary covering and discard it. Prepare the deck and apply permanent roof.

Over Steel Decks

First: Apply ¾" minimum thickness of Fesco Board insulation to provide a reasonable working surface, using cold adhesive in strips about 6" on center.

Second: Apply one layer of coated base felt solidly mopped with 190° asphalt, lapping the courses 4" over the preceding one.

Third: For extra protection mop on a surfacing of 190° asphalt.

Fourth: When the permanent roof is to be applied, tear off all the insulation and temporary covering and discard it. Prepare the deck and apply new insulation and the permanent roof.

Johns-Manville
Asbestogard
VAPOR RETARDANT SYSTEM

for use with FESCO® BOARD or FESCO-FOAM™ roof insulation over METAL DECKS

ASBESTOGARD ADHESIVE

ASBESTOGARD FELT

Specification **No. AG-1** *for use with*

FESCO® BOARD AND FESCO-FOAM ROOF INSULATION AND BUILT UP ROOFS OVER METAL DECKS

on inclines of up to ***6"*** *per foot*

Johns-Manville

Asbestogard VAPOR RETARDANT SYSTEM

tested and approved for Class I Insulated Metal Roof Deck Construction by the Factory Mutual Engineering Association laboratories.

Note: All information contained in "General Instructions" in the current Specification Manual for Johns-Manville Built-Up Roofs shall be considered part of this specification.

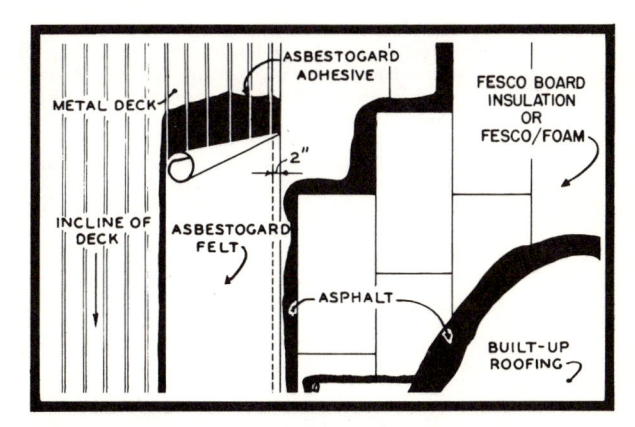

Application Instructions

First: Using a roller apply Asbestogard Adhesive in a continuous film to the top of the metal deck and parallel to the rib openings. Use 0.4 gallons of adhesive per square.

Apply Asbestogard Felt parallel to the ribs of the metal deck, lapping each sheet at least 2" at the sides and 4" at the ends. Form all side laps over solid bearing. Seal all laps with Asbestogard Adhesive.

The Asbestogard Felt shall be turned up on, but not cemented to all vertical surfaces to a height 4" to 6" and shall overhang all roof edges a similar amount.

Second: Over the Asbestogard Felt apply the units of Fesco Board or Fesco-Foam Insulation with long joints continuous and end joints staggered. Mop the full width under each unit of insulation with J-M Asphalt at a rate of approximately 25 lbs. per square. Form continuous long joints perpendicular to rib openings and with solid bearing.

If practical, it is recommended that the insulation be applied in more than one layer. Succeeding layers shall be applied in the same manner as the first layer, except 30 lbs. of Asphalt per square shall be used. All joints shall be staggered between layers.

Before application of the roofing, the projecting 6" of Asbestogard felt at vertical surfaces and at all edges shall be turned over the insulation and mopped solidly with the asphalt.

The insulation shall not be left exposed to the weather. No more insulation shall be applied than can be completely covered with the finished built-up roofing on the same day.

Nailing Strips

On decks where the incline is such that nailing of roofing felts is required, 2" and over for smooth surfaced roofs and 1" and over for gravel surfaced roofs, wood nailing strips shall be provided at ridge and at intermediate points not exceeding 20'-0" centers. Nailing strips the same thickness as the insulation shall be run horizontally to receive the insulation and retain nails securing the felts. On decks where the incline is 3" per foot and over, nailing strips shall be installed 4'-0 ¼" from inside face to inside face, to receive insulation and retain nails securing the felts.

Nailing strips and wood edging or curbs shall be of treated wood by the pressure process, with a water-borne salt, approved by the Wood Preserver's Association. Oil based preservatives such as creosote are not acceptable as they are not compatible with Asphalt Roofing components.

Bitumen

On slopes up to 3" per foot use J-M 190 Asphalt and on slopes 3" to 6" per foot J-M 220 Asphalt should be used.

MINIMUM THICKNESS OF FESCO BOARD OR FESCO FOAM OVER METAL DECKS			
Width of Rib Opening	Up to 1"	Up to 1¾"	Up to 2½"
Thickness of Insulation	¾"	1"	1½"

Notes:

1. If application of Asbestogard System and Fesco Board or Fesco-Foam insulation is to meet Factory Mutual Class I Construction, details of current FM I-28 should be followed.
2. To apply Asbestogard Adhesive use a roller 18" to 24" wide with ¾" to 1" nap. Pour adhesive into a deep pan or asphalt "buggy" wide enough to accommodate roller.
3. If Asbestogard Felt is damaged under foot traffic it can be readily patched with Asbestogard Felt and Adhesive.

BUILT-UP ROOFS by Johns-Manville/Greenwood Plaza, Denver, Colorado 80217

WATERPROOFING & DAMP PROOFING

Johns-Manville
Membrane Waterproofing System
For Exterior Walls and Floors Below Grade

*Specification **No. WP-1** for use over*

MASONRY SURFACES

Johns-Manville
Membrane Waterproofing System
For exterior walls and floors below grade

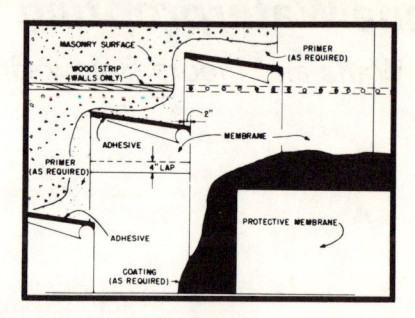

Preparation for Waterproofing
All surfaces shall be smooth, dry and firm with all cracks or voids filled with cement mortar. Concrete shall be smoothed to a float finish.

Nailing Strip
If wall is not nailable, a wood nailing strip must be provided, by other than the waterproofing contractor, set flush with the face of the wall, 6" above grade.

Primer
Coat the masonry surface over which the waterproofing is to be applied with the specified primer.

Application of Membrane
Coat with adhesive the area to receive the membrane. Broom the membrane into place, lapping long edges of each sheet 2". End laps shall be 4". Secure each layer at the top of the wall by nailing.

Install alternate layers of membrane and coatings of adhesive as called for by job conditions, allowing each layer to dry thoroughly before application of succeeding layers. After all layers of adhesive and membrane have been applied, coat the entire membrane with an additional coating of adhesive if specified.

Reinforcement
At all angles, projections or changes in plane in the wall or floor install a reinforcing of membrane strips, the first 12" wide covered by a second strip 18" wide, both set in and covered by a coating of adhesive if required. These reinforcing strips are to be applied over the completed waterproofing membrane.

If waterproofing is to connect with that of intersecting surfaces, provide a 12" lap in all layers, left dry, until woven into layers of intersecting waterproofing. These laps shall be protected and kept dry and clean.

Protection
Floor membrane must be immediately protected by the application of cement mortar or the temporary application of a layer of Planet Base Felt. In planting areas a protective layer of ¼" Flexboard may be installed over the completed membrane instead of cement mortar.

Wall membrane must be protected by the application of cement mortar or other masonry material. During back-filling, membrane must be protected by the installation of J-M Flexboard braced to hold in place until the back-fill is completed.

Approximate Quantities of Materials Required for 100 Sq. Ft.

Materials are for one layer of waterproofing only. Additional quantities must be added according to hydrostatic conditions.

	MEMBRANE	PRIMER	ADHESIVE	COATING
HOT APPLIED	No. 15 Asbestos Waterproofing Felt ... 15 lbs.	Concrete Primer ... 1 gal.	190 Asphalt ... 20 lbs.	190 Asphalt ... 20 lbs.
COLD APPLIED	Cold Appl'n Asb. Felt ... 35 lbs.	Concrete Primer ... 1 gal.	Cold Application Cement 2 gal.	None

Hydrostatic Conditions Which Determine Waterproofing Requirements

HYDROSTATIC HEAD IN FEET →			1	2	3	4	5	6	7	8	9	10	12
	PRESSURE IN LBS. PER SQ. IN.		0.43	0.86	1.30	1.73	2.17	2.60	3.01	3.47	3.87	4.34	5.21
NUMBER OF LAYERS OF WATERPROOFING MATERIALS	HOT APPLICATION • ASBESTOS FELT and ASPHALT	PRIMER	1	1	1	1	1	1	1	1	1	1	1
		FELT	1	1	1	2	2	2	2	2	2	3	3
		ASPHALT	2	2	2	3	3	3	3	3	3	4	4
	COLD APPLICATION • COLD APPLICATION ASBESTOS FELT	PRIMER	1	1	1	1	1	1	1	1	1	1	1
		FELT	1	1	1	1	1	1	2	2	2	2	2
		CEMENT	1	1	1	1	1	1	2	2	2	2	2
FLOOR PROTECTION	CEMENT MORTAR		1"	1"	1"	1"	1"	1½"	1½"	1½"	1½"	2"	2"
	LAYER OF PLANET BASE FELT		1	1	1	1	1	1	1	1	1	1	1
WALL PROTECTION	CEMENT MORTAR		1½"	1½"	1½"	1½"	1½"	2"	2"	2"	2"	2½"	2½"
	BRICK or TERRA COTTA		4"	4"	4"	4"	4"	4"	4"	4"	4"	4"	4"
	LAYER OF FLEXBOARD		1	1	1	1	1	1	1	1	1	1	1

J-M NO. 15 Asbestos Waterproofing Felt ... an asphalt-saturated asbestos felt.

J-M NO. 190 Asphalt ... a high grade asphalt so named for its melting point of approximately 190 F.

J-M Cold Application Asbestos Felt ... an asphalt saturated, double coated Asbestos Felt for cold application.

J-M Cold Application Cement ... a heavy brushing, cut-back asphalt for applying cold application felts.

J-M Planet Base Felt ... a heavy rag felt, saturated and coated on both sides with asphalt.

J-M Flexboard ... an asbestos-cement sheet used for a wide range of exterior and interior applications.

Diagrams

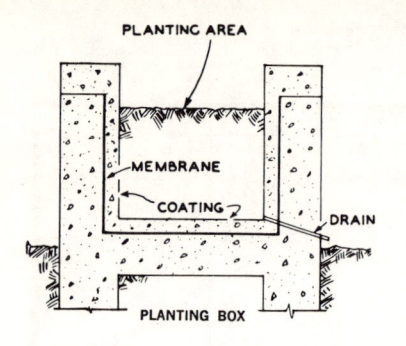

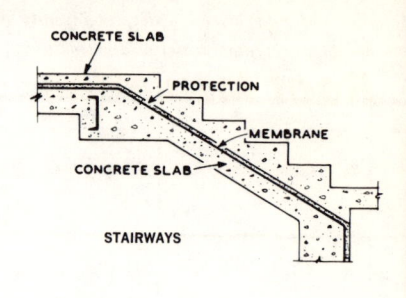

PLANTING BOX — labels: PLANTING AREA, MEMBRANE, COATING, DRAIN

STAIRWAYS — labels: CONCRETE SLAB, PROTECTION, MEMBRANE, CONCRETE SLAB

OUTSIDE WALL MEMBRANE (Outside Application) — labels: PRIMER, MEMBRANE, PROTECTION, CONCRETE WALL, FLOOR SLAB, MEMBRANE, PRIMER, BASE SLAB, FOOTING

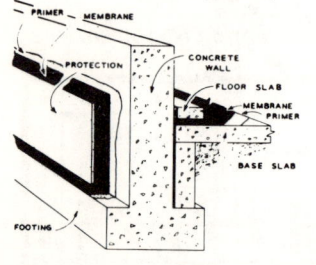

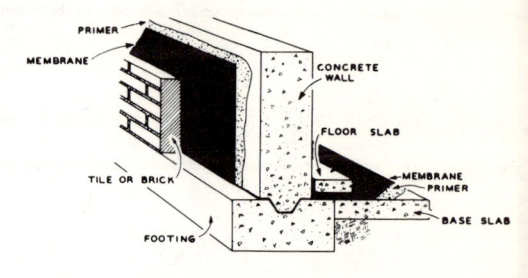

(Upper right diagram) — labels: PRIMER, MEMBRANE, CONCRETE WALL, FLOOR SLAB, TILE OR BRICK, MEMBRANE, PRIMER, BASE SLAB, FOOTING

OUTSIDE WALL MEMBRANE (Inside Application) — labels: MEMBRANE, TILE OR BRICK, CONCRETE WALL, FLOOR SLAB, PRIMER, KEY, BASE SLAB, FOOTING

IMPORTANT

Recommendations for the use of these products are based upon experience and tests believed to be reliable. However, since the use and/or application of these systems are beyond the control of Johns-Manville, Johns-Manville can assume no responsibility therefor. All warranties, guarantees, obligations or promises expressed or implied by contract or by law are expressly disclaimed.

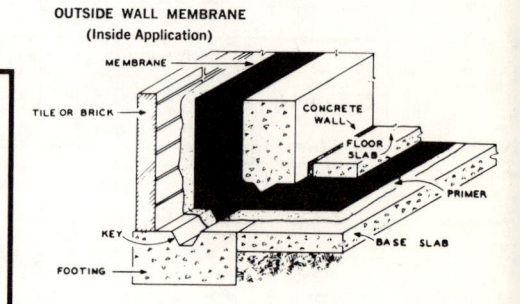

BUILT-UP ROOFS by Johns-Manville/Greenwood Plaza, Denver, Colorado 80217

BU-224A-WP-1 1/76

Litho in U.S

Roof Insulation

*Specification **No. 500** for use over*

NAILABLE DECKS

*on inclines up to **6"** per foot*

Johns-Manville
Fesco Board and Fesco-Foam
ROOF INSULATION FOR BUILT-UP ROOFS

This specification is for use over any type of structural deck which can receive and adequately retain nails or other types of mechanical fasteners recommended by the deck manufacturer.

Note: All information contained in "General Instructions" in the current Specification Manual for Johns-Manville Built-Up Roofs shall be considered part of this specification.

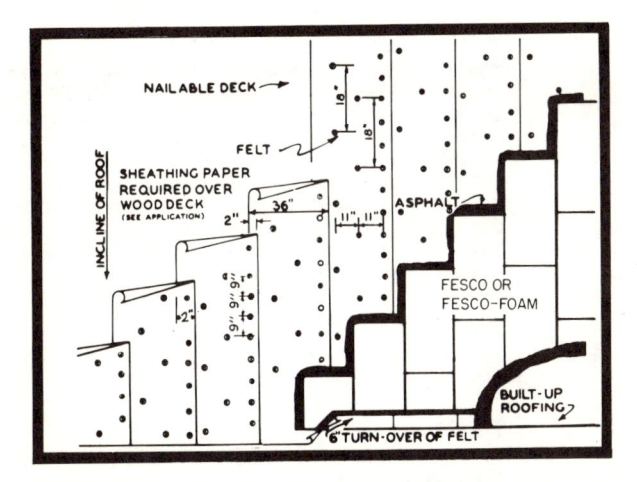

Application of Insulation

If application is over wood boards, cover the deck with one layer of sheathing paper lapping each layer not less than 1" and nail sufficiently to hold in place.

First: If a felt is required apply the one layer of felt lapping each felt 2" over the preceding one and nail through the laps at 9" centers and down the longitudinal center of each felt two rows of nails with the nails spaced approximately 11" apart and nails staggered on approximately 18" centers. Use nails or fasteners appropriate to the type deck. This felt shall be turned up on, but not cemented to, all vertical surfaces to a height of 6" and shall overhang all roof edges a similar amount.

If application is over plywood sheathing paper may be omitted and felt may be sprinkle mopped to the deck using 10 lbs. of asphalt per square, in Region 3 only.

Second: Apply the units of insulation with the long joints continuous. Short joints shall be staggered. (over wood boards the long edge of units are placed at right angles to the boards).

Fesco-Foam must be applied with the felt side up.

Mop the full width under each unit of insulation with hot Asphalt at a minimum rate of 30 lbs per square per layer of insulation.

Third: Before the application of roofing, the projecting 6" of felt at vertical surfaces and at all edges shall be turned over the insulation and mopped solidly with the asphalt.

When Fesco is installed without a felt, each unit of insulation shall be nailed on 24" centers adjacent to the long edges using 6 nails. Fesco-Foam shall be nailed in a similar manner except an additional row of nails shall be added down the longitudinal center with nails staggered using 8 nails.

Special Instructions

The insulation shall not be left exposed to the weather. No more insulation shall be applied than can be completely covered with the finished built-up roofing on the same day.

Where the incline of the roof is such, that nailing of the roofing felts is required, 2" and over for smooth surface roofs and 1" and over for gravel surface roofs, each unit of Fesco Board shall be nailed on 24" centers adjacent to the long edges. Fesco-Foam shall be secured with an additional row of nails and down the longitudinal center with the nails staggered. If applied in more than one layer all nailing shall be through the top layer. If practical, it is recommended that the insulation be applied in more than one layer, staggering joints between layers and mopping between layers with 30 lbs of Asphalt per square, nailing through the top layer.

Nailing

All nails or other fasteners are to be driven through tin caps unless the nail or fastener has an integral flat cap no less than 1" across.

Felts

One layer of J-M Asbestos Finishing Felt or J-M Asphalt-Saturated Felt shall be used.

Bitumen —On slopes up to 3" per foot use J-M 190 Asphalt and on slopes 3" to 6" per foot J-M 220 Asphalt should be used.

BUILT-UP ROOFS by Johns-Manville/Greenwood Plaza, Denver, Colorado 80217

Specification **No. 501** *for use over*

NON-NAILABLE DECKS

*on inclines up to **6"** per foot*

Johns-Manville
Fesco Board and Fesco-Foam
ROOF INSULATION FOR BUILT-UP ROOFS

General — This specification is for use over any type of structural deck which is not nailable and which offers a suitable surface to receive the insulation. Poured and pre-cast concrete decks require priming.

Note: All information contained in "General Instructions" in the current Specification Manual for Johns-Manville Built-Up Roofs shall be considered part of this specification.

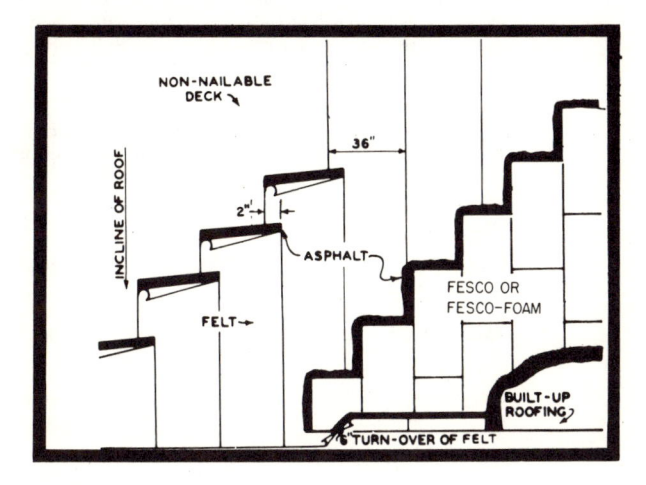

Application of Insulation

First: If a felt is required apply the one layer of felt lapping each felt 2" over the preceding one and mopping the full width under each felt with J-M Asphalt at a minimum rate of 23 lbs per square. Edges of the felt shall be turned up on, but not cemented to, all vertical surfaces to a height of 6" and shall overhang all roof edges a similar amount.

Second: Apply the units of insulation with long joints continuous and short joints staggered.

Fesco-Foam must be applied with the felt side up.

Mop the full width under each unit of insulation with J-M Asphalt at a minimum rate of 30 lbs per square per layer of insulation.

Special Instructions

If practical, it is recommended that the insulation be applied in more than one layer. Succeeding layers shall be applied in the same manner as the first layer. All joints shall be staggered between layers.

Before application of the roofing, the projecting 6" of felt at vertical surfaces and at all edges shall be turned over the insulation and mopped solidly with the asphalt. The insulation shall not be left exposed to the weather. No more insulation shall be applied than can be completely covered with the finished built-up roofing on the same day.

Units of insulation can be installed directly to some decks in a solid mopping of asphalt using 23 lbs. per square.

NAILING STRIPS

On non-nailable decks where the incline is such that nailing of roofing felts is required, 2" and over for smooth surfaced roofs and 1" and over for gravel surfaced roofs, wood nailing strips shall be provided at ridge and at intermediate points not exceeding 20'-0" centers. Nailing strips, the same thickness as the insulation, shall be run horizontally to receive the insulation and retain nails securing the felts. On insulated roof decks where the incline is 3" per foot and over, nailing strips shall be installed 4'-0 1/4" from inside face to inside face, to receive the insulation and retain nails securing the felt.

Nailing strips and wood edging or curbs shall be of treated wood by the pressure process with a water borne salt as approved by the American Wood Preserver's Assn. Oil based preservatives such as creosote are not acceptable as they are not compatible with Asphalt Roofing Components.

FELTS

One layer of J-M Asbestos Finishing Felt or J-M Asphalt-Saturated Felt shall be used.

Bitumen—On slopes up to 3" per foot use J-M 190 Asphalt and on slopes 3" to 6" per foot J-M 220 Asphalt should be used.

BUILT-UP ROOFS by Johns-Manville/Greenwood Plaza, Denver, Colorado 80217

Specification **No. 502** for use over
STEEL DECKS
*on inclines up to **6"** per foot*

Johns-Manville
Fesco Board and Fesco-Foam
ROOF INSULATION FOR BUILT-UP ROOFS

Preparation of Deck — Steel deck, minimum 22 gauge, shall be dry, clean and properly graded to all outlets. A wood strip of the same thickness as the insulation shall be secured to the roof deck adjoining all eaves to act as a stop for the insulation.

Note: All information contained in "General Instructions" in the current Specification Manual for Johns-Manville Built-Up Roofs shall be considered part of this specification.

If the steel deck surface is such that a satisfactory bond cannot be obtained between the deck and the adhesive, coat the deck with Concrete Primer and allow to dry.

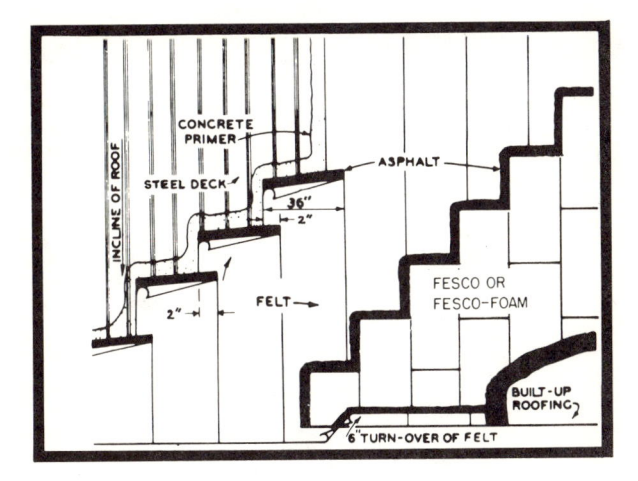

Application of Insulation

First: Apply the one layer of felt lapping each felt 2" over the preceding one and mopping the full width under each felt with J-M Asphalt at a minimum rate of 23 lbs per square. Edges of the felt shall be turned up on, but not cemented to, all vertical surfaces to a height of 6" and shall overhang all roof edges a similar amount.

Second: Apply the units of insulation with long joints continuous either parallel or at right angles to the ribs of the metal deck. When joints are parallel to ribs they must be formed over solid bearing. All end joints shall be staggered.

Mop the full width under each unit of insulation with J-M Asphalt at a minimum rate of 30 lbs per square per layer of insulation.

Insulation can be applied directly to steel decks using either 12-25 lbs of hot asphalt applied in ribbons 6" O.C. or an approved cold adhesive using 0.4-0.8 gal. per square in ribbons 6" O.C.

Mechanical fasteners can be used in addition to or instead of adhesive methods for securement provided 4 approved fasteners are used per unit of Fesco & 6 approved fasteners are used per unit of Fesco-Foam.

Special Instructions

If practical, it is recommended that the insulation be applied in more than one layer. Succeeding layers shall be applied in the same manner as the first layer. All joints shall be staggered between layers.

Fesco-Foam must be applied with the felt side up.

Before application of the roofing, the projecting 6" of felt at vertical surfaces and at all edges shall be turned over the insulation and mopped solidly with the asphalt.

The insulation shall not be left exposed to the weather. No more insulation shall be applied than can be completely covered with the finished built-up roofing on the same day.

Nailing Strips

On decks where the incline is such that nailing of roofing felts is required, 2" and over for smooth surfaced roofs and 1" and over for gravel surfaced roofs, wood nailing strips shall be provided at ridge and at intermediate points not exceeding 20'-0" centers. Nailing strips the same thickness as the insulation shall be run horizontally to receive the insulation and retain nails securing the felts. On decks where the incline is 3" per foot and over, nailing strips shall be installed 4'-0-¼" from inside face to inside face, to receive insulation and retain nails securing the felts.

Nailing strips and wood edging or curbs shall be of treated wood by the pressure process with a water borne salt as approved by the American Wood Preserver's Assn. Oil based preservatives such as creosote are not acceptable as they are not compatible with Asphalt Roofing Components.

Felts

One layer of J-M Asbestos Finishing Felt, J-M Asphalt-Saturated Felt or the J-M Asbestogard System shall be used.

Bitumen —On slopes up to 3" per foot use J-M 190 Asphalt and on slopes 3" to 6" per foot J-M 220 Asphalt should be used.

Note: If application of insulation must comply with either Factory Mutual Class I Construction or Underwriters Laboratories' Construction No. 1 or No. 2 follow details as set forth in current Factory Mutual Construction I-28, Underwriters Laboratories Building Materials List, or Johns-Manville Asbestogard System.

Minimum Thickness of Fesco or Fesco-Foam over Metal Decks

Width of Rib Opening	Up to 1"	Up to 1¾"	Up to 2½"
Thickness of Insulation	¾"	1"	1½"

BUILT-UP ROOFS by Johns-Manville/Greenwood Plaza, Denver, Colorado 80217